SpringerBriefs in Evolutionary Biology

SpringerBriefs present concise summaries of cutting-edge research and practical applications across a wide spectrum of fields. Featuring compact volumes of 50 to 125 pages, the series covers a range of content from professional to academic. Typical topics might include:

- A timely report of state-of-the art analytical techniques
- A bridge between new research results, as published in journal articles, and a contextual literature review
- A snapshot of a hot or emerging topic
- An in-depth case study
- A presentation of core concepts that students must understand in order to make independent contributions

Rodrick Wallace

Essays on the Extended Evolutionary Synthesis

Formalizations and Expansions

 Springer

Rodrick Wallace
Division of Epidemiology
The New York State Psychiatric Institute
New York, NY, USA

ISSN 2192-8134   ISSN 2192-8142   (electronic)
SpringerBriefs in Evolutionary Biology
ISBN 978-3-031-29878-3   ISBN 978-3-031-29879-0   (eBook)
https://doi.org/10.1007/978-3-031-29879-0

This Springer imprint is published by the registered company Springer Nature Switzerland AG
The registered company address is: Gewerbestrasse 11, 6330 Cham, Switzerland

Preface

> *Different representations of evolutionary theory can be considered analogous to the tools in a toolbox. There is no question of whether a hammer is better than a wrench in general; only whether one is more appropriate to the task at hand. Therefore, an assessment of the fruitfulness of a representation of evolutionary theory will be keyed to the task that motivates the inquiry.*
>
> — Love (2010)

From the 'punctuated equilibrium' of Eldrege and Gould (Gould 2002, Ch. 9) through Lewontin's 'triple helix' (Lewontin 2002) and the various visions and revisions of the Extended Evolutionary Synthesis (EES) of Pigliucce and Muller 10 (2010), Laland et al. (2014), and EES (2020), both data and theory have demanded an opening-up of the 1950s Modern Synthesis that so firmly wedded evolutionary theory to the mathematics of gene frequency analysis. Gould (2002) in particular carries much of the historiographic burden.

It can be argued that a single deep and comprehensive mathematical theory may simply not be possible for the almost infinite varieties of evolutionary process active at and across the full range of scales of biological, social, institutional, and cultural phenomena. Indeed, the spectacular failure of 'meme theory' should have raised a red flag that narrow gene-centered models of evolutionary process may indeed have serious limitations (Chvaja 2020).

What is attempted in this work is somewhat less grand, but still broad. Following the instruction of Maturana and Varela (1980) that all living systems are cognitive, in a certain sense, and that living as a process is a process of cognition, the asymptotic limit theorems of information and control theories that bound all cognition (Dretske 24 1994) provide a basis for constructing an only modestly deep but wider-ranging series of probability models that might be converted into useful statistical tools for the analysis of observational and experimental data related to evolutionary process. Here, in a series of interrelated essays, we attempt to outline such a theory. This proves to be surprisingly direct.

The first chapter, from Wallace (2022a), explores the major transitions from a novel perspective, based on the breaking of groupoid symmetries associated with information processes, and provides a foundation for understanding the later chapters.

Chapter 2 focuses on an expansion of the EES based on information and control theories, while Chap. 3 is dedicated specifically to the dynamics of regulation, leading, in Chap. 4, to a generalization of Casanova and Konkel's (2020) elegant work beyond their genetic focus. Chapter 5 makes application to the evolution of institutions, expanding work by Wallace and Fullilove (2014) and Wallace (2013). Chapter 6 explores 'speciation' fragmentation, while Chap. 7 makes application to recent work on biomolecular condensate dynamics. The final chapter, adapted from Wallace (2022c) examines the evolutionary exaptation of crosstalk 'leakage' to study shared interbrain activity in social communication.

The chapters, while broadly interrelated, can be read separately.

Conversion of this theory into validated statistical tools, however, remains to be done and will not be trivial.

We begin with a brief methodological overview.

An Introduction to the Formalism

Before proceeding to details, it is useful to review something of the 'fundamental homologies' across statistical thermodynamics, information theory, and control theory.

Recall Einstein's treatment of Brownian motion for N particles, the first of the 'fluctuation-dissipation' studies.

The simple diffusion equation

$$\partial \rho(x, t)/\partial t = \mu \partial^2 \rho(x, t)/\partial x^2 \tag{1}$$

has a solution in terms of the Normal Distribution as

$$\rho(x, t) = \frac{N}{\sqrt{4\pi \mu t}} \exp[-x^2/(4\mu t)] \tag{2}$$

A little calculation gives

$$\sqrt{<x^2>} \propto \sqrt{t} \tag{3}$$

Feynman (2000), following Bennett, shows how a 'simple ideal machine' permits extraction of free energy from an information source. The Rate Distortion Theorem (Cover and Thomas 2006) expresses the minimum channel capacity $R(D)$ needed for a transmitted signal along a noisy channel to be received with average distortion less than or equal to some scalar measure D. The 'worst-case scenario' is the Gaussian channel for which, under a square distortion measure,

$$R(D) = \frac{1}{2}\log_2(\sigma^2/D), \quad D \le \sigma^2$$

$$R(D) = 0, \quad D > \sigma^2 \tag{4}$$

If—following Feynman (2000)—we identify information as a form of free energy, we can also define a classic thermodynamic entropy via the Legendre transform as

$$S \equiv -R(D) + D dR/dD \tag{5}$$

The next step imposes a first-order nonequilibrium thermodynamics Onsager approximation (de Groot and Mazur 1984) as

$$dD/dt \propto dS/dD = D d^2 R/dD^2$$

$$dD/dt = \frac{1}{2\log(2)D(t)}$$

$$D(t) \propto \sqrt{t} \tag{6}$$

In the absence of 'control', in a large sense, distortion simply grows as a diffusion, for this worst-case example.

This is a basic and somewhat surprising result, suggesting validity to the generalized Onsager nonequilibrium formalism used in what follows. Indeed, the method provides the basis for a heuristic derivation of the relation between information and control theories, the Data Rate Theorem (Nair et al. 2007). That theorem characterizes the minimum control information $\mathcal{H}$ needed to stabilize an inherently unstable control system.

Suppose we impose 'control free energy' at some rate $M(\mathcal{H})$ on an inherently unstable system diffusing from control as $D(t) \propto \sqrt{t}$. A necessary condition for stability can be derived as

$$dD/dt = \mu D d^2 R/dD^2 - M(\mathcal{H}) \le 0$$

$$M(\mathcal{H}) \ge \mu D d^2 R/dD^2 \ge 0$$

$$\mathcal{H} \ge \mathcal{H}_0 \equiv \max\{M^{-1}(\mu D d^2 R/dD^2)\}\} \tag{7}$$

where max represents the maximum value and noting that if M is monotonic increasing, so is the inverse function M^{-1}. By convexity, $d^2 R/dD^2 \ge 0$ (Cover and Thomas 2006).

For the Gaussian channel—in the absence of further noise—at nonequilibrium steady state, where $dD/dt \equiv 0$, the average distortion can be expressed as

$$D \propto 1/M(\mathcal{H}) \tag{8}$$

Something similar emerges for the 'average-average' distortion $< D >$ of a Gaussian channel even under full generalization via the stochastic differential equation

$$dD_t = \left[\mu Dd^2 R/dD^2 - M(\mathcal{H}) \right] dt + \Sigma D_t dB_t \tag{9}$$

where, in the latter term, Σ is the magnitude of the 'volatility' under Brownian white noise dB_t. Such stochastic extensions, via the Ito Chain Rule, may impose additional stability constraints (e.g., Appleby et al. 2008). For example, using the Chain Rule, 'it is easy to show' that stability in variance of Eq. (9) for the volatility model based on Eq. (4) requires

$$M(\mathcal{H}) = D\frac{\Sigma^2}{2} + \frac{\mu}{D \log(4)}$$

$$M(\mathcal{H}) \geq \Sigma \sqrt{\frac{\mu}{\log(2)}}$$

$$\mathcal{H} \geq M^{-1}\left(\Sigma \sqrt{\frac{\mu}{\log(2)}} \right) \tag{10}$$

Other characterizations of stability are, of course, also possible.

The calculations illustrate the essence of the Data Rate Theorem connecting information and control theories.

On the Difference Between Cognitive and Physical System Dynamics

Overall, these results are in the direction of the elegant 'fluctuation-dissipation theorems' studied by Jarzynski (1997), Barato and Siefert (2015), Gingrich et al (2016), and many others. But there are essential differences.

We are assuming that cognitive systems are driven by the interaction between rates of available 'material' or 'materiel' free energy, internal communication bandwidth, and the rate at which 'intelligence' information is perceived from the external world, taking, for example, a scalar value Z as the product of those three rates. This is explored in later chapters, where more complicated examples are studied.

Recall the central argument regarding information process in nonergodic systems, where source uncertainty is path dependent, i.e., each high probability path has its own limiting value of source uncertainty, not, however, given as a 'Shannon entropy' (Khinchin 1957).

For each high probability path j, it is then possible to write a Boltzmann pseudoprobability as

$$P_j = \frac{\exp[-H_j/g(Z)]}{\sum_k \exp[-H_k/g(Z)]} \tag{11}$$

H_k is the source uncertainty of the high probability path k and the sum—or generalized integral—is over all high probability paths. $g(Z)$ is a scalar 'cognitive temperature' analog depending on the essential resource rate Z, assumed here to be a scalar. In subsequent chapters, $g(Z)$ will be calculated from first principles.

Again, Z may index available internal and/or external information bandwidths, rate of metabolic free energy in a physiological system, rate of 'materiel' and personnel supply in organized conflict, or some synergism of these that, like a principle component analysis, project down to one dimension but accounts for a substantial fraction of overall variance.

It is then possible to define a free energy Morse Function F (Pettini 2007) in terms of the 'partition function' denominator of Eq. (11),

$$\exp[-F/g(Z)] = \sum_k \exp[-H_k/g(Z)] \equiv h(g(Z))$$

$$F(Z) = -\log[h(g(Z))]g(Z)$$

$$g(Z) = -\frac{F(Z)}{RootOf\left(e^X - h\left(-\frac{F(Z)}{X}\right)\right)} \tag{12}$$

where X is a dummy variate whose solution gives the desired expression. Setting $h(g(Z)) = g(Z)$ shows the RootOf construct generalizes the Lambert W-function of order n that satisfies the relation $W(n,x)\exp[W(n,x)] = x$. In that case, $g(Z) = -F(Z)/W(n, -F(Z))$. (This can happen if the sum in Eq. (11) can be approximated by an integral in $\exp[-H/g(Z)]$ over the range $0 \rightarrow \infty$.) The Lambert W-function is real-valued only for $n = 0, -1$ over the respective ranges $x \geq -\exp[-1]$ and $-\exp[-1] \leq x \leq 0$.

The RootOf relation for $g(Z)$ may, in general, have complex solutions, representing phase transitions analogous to those of 'Fisher zeros' in physical systems (e.g. Dolan et al. 2001 and references therein).

Somewhat surprisingly, it is possible to contrast and compare such cognitive phase transitions with Jarzynski's (1997) classic derivation of a nonequilibrium equality for free energy differences.

The second law of thermodynamics famously states that for *finite time* transitions between equilibrium states along any path, the average work needed will be greater than or equal to the change in free energy ΔF between the states:

$$< \mathscr{W} > \geq \Delta F \tag{13}$$

where the '$< >$' bracket indicates an average over an ensemble of measurements of the work $\mathscr{W}$. Jarzynski's elegant extension is the *exact* statement

$$< \exp[-\mathscr{W}/kT] >= \exp[-\Delta F/kT]$$

$$\Delta F = -kT \log\left(< \exp[-\mathscr{W}/kT] >\right) \equiv -kT \log\left(h(kT)\right) \qquad (14)$$

where T is the absolute temperature and k an appropriate constant. Equation (13) emerges from the first of these expressions via the Jensen inequality $< \phi(x) > \geq \phi(< x >)$ for any convex function ϕ.

Inverting the perspective—thus making free energy more fundamental than temperature—*we can solve the Jarzynski relation for kT*, giving, as above, a 'Fisher zero' set of phase transition solutions as

$$kT = -\frac{\Delta F}{RootOf\left(e^X - h\left(-\frac{\Delta F}{X}\right)\right)} \qquad (15)$$

where, again, X is a dummy variate in the RootOf construct generalizing the Lambert W-function.

That function, in fact, emerges directly if $h(kT)$ were imagined to have a strongly dominant term $a(kT)^b$. Then

$$kT \approx (-\Delta F)/[b\, W(n, -\Delta F a^{1/b}/b)] \qquad (16)$$

Taking $n = 0, -1$, this expression will again be subject to 'Fisher zero' imaginary component phase transitions if ΔF exceeds appropriate ranges.

These results emphasize the considerable degree to which the dynamics of cognitive processes are likely to differ from the more familiar dynamics of physical—or even biophysical—phenomena. That is, cognitive systems are channeled by the asymptotic limit theorems of information and control theories (Dretske 1994), in addition to usual physical constraints, and this is most often a different world indeed.

More specifically, while, for physical systems, free energy dynamics may be dominated by 'ordinary' temperature changes across phase transitions, cognitive phase transitions—involving the 'cognitive temperature' $g(Z)$ of Eqs. (11) and (12)—normally occur at fixed physical temperature and are *driven by changes in rates of available free energies that include rates of information transmission*.

Singular characteristics of cognitive dynamics, in particular 'Yerkes-Dodson laws' and their generalizations, emerge from this fundamental difference.

A further divergence from physical theory emerges in our adaptation of Onsager's formalism for nonequilibrium thermodynamics, defining a free energy analog using the 'partition function' of the denominator of Eq. (11). This permits construction of an entropy-analog via a Legendre transform, and versions of the basic Onsager approach, i.e. time derivatives of driving parameters being taken as proportional to the divergence of entropy by those parameters, extending Eqs. (5) and (6). Multi-dimensional forms of these models, however, do not sustain 'Onsager reciprocal relations' since information sources are not microreversible: in English 'eht' does not have the same probability as 'the'. Indeed, for information sources in general, palindromes have vanishingly small probabilities.

That is, the topology of information sources is dominated by directed homotopy, leading to equivalence class groupoid symmetries and phase transitions in terms of broken groupoid symmetries. Indeed Cimmelli et al. (2014) describe similar matters afflicting nonlinear nanoscale phenomena that appear consonant with the fluctuation-dissipation studies cited above.

Iterating an Approach

We can, from a control perspective in which a scalar distortion measures the difference between what is ordered and what is observed, use this development to iterate the Rate Distortion Function argument. That is, rather than viewing $R(D)$ itself as 'the' free energy, a more complicated system will use R as in Eq. (11), defining a pseudoprobability in Z as

$$dP = \frac{\exp[-R/g(Z)]dR}{\int_0^\infty \exp[-R/g(Z)]dR} = \frac{\exp[-R/g(Z)]dR}{g(Z)} \tag{17}$$

leading to definition of a free energy F from the partition function denominator of Eq. (17) as

$$\exp[-F/g(Z)] = g(Z)$$
$$g(Z) = \frac{-F}{W(n, -F)} \tag{18}$$

where, again, $W(n, x)$ is the Lambert W-function of order n.

Dynamics again emerge according to an Onsager approximation in the gradient of an entropy-analog as

$$S \equiv -F(Z) + ZdF/dZ$$
$$dZ/dt \approx \mu dS/dZ = Zd^2F/dZ^2 = f(Z) \tag{19}$$

The approach of Eq. (11) et seq., however, often permits a more textured analysis of individual systems.

Waiting for Kepler: Inadequacy of the Ergodic Decomposition

Here, we follow closely the arguments of Wallace (2022b).

According to popular mathematical lore, nonergodic information sources are a 'trivial matter' as a consequence of the Ergodic Decomposition Theorem (Gray 1988 Ch.7) which states that it is possible to factor any nonergodic process into a sum or generalized integral of individual ergodic processes, in the same way that any point on a triangle can be expressed in terms of its extremal fixed point vertexes. According to Winkelbauer (1970),

Theorem II *The asymptotic rate of a stationary source μ equals the essential supremum of the entropy rates of its ergodic components:*

$$H(\mu) = ess. \sup_{z \in R[\mu]} H(\mu_z)$$

where the μ_z are ergodic.

This is, however, not a 'simple' result for dynamic systems that can suffer 'absorbing states': individual paths—and small, closely related, equivalence classes of them—are critically important in biological phenomena because each path may have a unique consequence for the organism or other cognitive entity embedded in and interacting with a stressful environment. For example, there will only be a single 'meaningful sequence' associated with successful capture by a predator. That is, absorbing states are particularly important in biological processes.

Recall, further, some singular intellectual history: it is possible to approximate any reasonably well-behaved real-valued function over a fixed interval in terms of a Fourier series. Recall that it was, in the geocentric Ptolemaic system, via a sufficient number of epicycles, possible to predict planetary positions to any desired accuracy using such a de facto Fourier Decomposition. The underlying astronomical problem was both considerably simplified and greatly enhanced by the non-geocentric empirical observations of Kepler, explained by Newton, and fully elaborated by Einstein.

The phenomena of cognition, consciousness, and evolution are considerably more complex than the motion of the planets around the sun, and Keplerian laws must still be found across many physiological, social, and institutional phenomena, and the like. Newton and Einstein are nowhere on the horizon for theories of cognition, the dynamics of automata, individual consciousness, evolutionary process, and collective analogs.

New York, NY, USA   Rodrick Wallace

References

Appleby, J., X. Mao, and A. Rodkina. 2008. Stabilization and destabilization of nonlinear differential equations by noise. IEEE Transactions on Automatic Control 53:68–69.

Barato, A., and U. Seifert. 2025. Thermodynamic uncertainty relation for biomolectular processes. Physical Review Letters 114:158101.

Casanova, E., and M. Konkel. 2020. The developmental gene hypothesis for punctuated equilibrium: combined roles of developmental regulatory genes and transposable elements. Bioessays 42:e1900173.

Chvaja, R. 2020. Why did memetics fail? Comparative case study. Perspectives in Science 28:542–570.

Cimmelli, V., A. Sellitto, and D. Jou. 2014. A nonlinear thermodynamic model for a breakdown of the Onsager symmetry and the efficiency of thermoelectric conversion in nanowires. Proceedings of the Royal Society A 470:20140265.

Cover, T., and J. Thomas. 2006. *Elements of Information Theory*, 2nd ed. New York: Wiley.

de Groot, S., and P. Mazur. (1984). *Nonequilibrium Thermodynamics*. New York: Dover.

Dolan, B., W. Janke, D. Johnston, and M. Stathakopoulos. 2001. Thin fisher zeros. Journal of Physics A 34:6211–6223.

Dretske, F. 1994. The explanatory role of information. Philosophical Transactions of the Royal Society A 349:59–70.

EES. 2020. https://extendedevolutionarysynthesis.com/resources/recommended-reading/

Feynman, R. 2000. *Lectures on Computation*. New York: Westview Press.

Gingrich, T., J. Horowitz, N. Peunov, and J. England. 2016. Dissipation bounds all steady-state current fluctuations. Physical Review Letters 116:120601.

Gould, S.J. 2002. *The Structure of Evolutionary Theory*. Cambridge: Harvard University Press.

Gray, R. 1988. Probability, *Random Processes, and Ergodic Properties*. New York: Springer.

Jarzynski, C. 1997. Nonequilibrium equality for free energy differences. Physical Review Letters 78:2690–2693.

Khinchin, A. 1957. *Mathematical Foundations of Information Theory*. New York: Dover.

Laland, K., T. Uller, M. Feldman, K. Sterelny, G. Muller, A. Moczek, E. Jablonka, AND J. Odling-Smee. 2014. Does evolutionary theory need a rethink? Nature 514:163–164.

Lewontin, R. 2002. *The Triple Helix: Gene, Organism and Environment*. Cambridge: Harvard University Press.

Love, A. 2010. Rethinking the structure of evolutionary theory for an extended synthesis. In *Evolution: The Extended Synthesis*, ed. Massimo Pigliucci and Gerd B. Müller (eds.), Chap. 16. Cambridge: MIT Press.

Maturana, H., and F. Varela. 1980. *Autopoiesis and Cognition: The Realization of the Living*. Boston: Reidel.

Nair, G., F. Fagnani, S. Zampieri, and R. Evans. 2007. Feedback control under data rate constraints: an overview. Proceedings of the IEEE 95:108138.

Pettini, M. 2007. *Geometry and Topology in Hamiltonian Dynamics and Statistical Mechanics*. New York: Springer.

Pigliucce, M., and G. Muller. 2010. *Evolution: The Extended Synthesis*. Cambridge: MIT Press.

R. Wallace. 2013. A new formal approach to evolutionary process in socioeconomic systems. Journal of Evolutionary Economics 23:1–15.

Wallace, R. 2022a. Major transitions as groupoid symmetry-breaking in nonergodic prebiotic, biological and social information systems. Acta Biotheoretica 70:27. https://doi.org/10.1007/s10441-022-09451-5

Wallace, R. 2022b. Consciousness, Cognition and Crosstalk: The Evolutionary Exaptation of Nonergodic Groupoid Symmetry-Breaking. New York: Springer.

Wallace, R. 2022c. Formal perspectives on shared interbrain activity in social communication: insights from informatin and control theories. Cognitive Neurodynamics. https://doi.org/10.1007/s11571-022-09811-4

Wallace, R., and M. Fullilove. 2014. State policy and the political economy of criminal enterprise: mass incarceration and persistent organized hyperviolence in the USA. Structural Change and Economic Dynamics 31:17–31.

Winkelbauer, K. 1970. On the asymptotic rate of non-ergodic information sources. Kybernetika Cislo 2. Rocnik 6/1970:127–148.

Contents

About the Author

Rodrick Wallace is a research scientist in the Division of Epidemiology at the New York State Psychiatric Institute, affiliated with the Columbia University Department of Psychiatry. He has an undergraduate degree in mathematics and a PhD in physics from Columbia, and completed postdoctoral training in the epidemiology of mental disorders at Rutgers. He has worked as a public interest lobbyist, conducting empirical studies of fire service deployment, and received an Investigator Award in Health Policy Research from the Robert Wood Johnson Foundation. In addition to material on public health and public policy, he has authored peer reviewed studies modeling evolutionary process and heterodox economics, as well as quantitative analyses of institutional and machine cognition. He publishes in the military science literature, and received one of the UK MoD RUSI Trench Gascoigne Essay Awards.

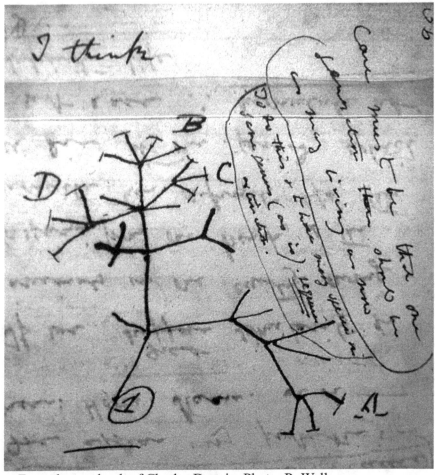

From the notebook of Charles Darwin. Photo: R. Wallace.

Chapter 1
On the Major Transitions

> *Living systems are cognitive systems, and living as a process is a process of cognition. This statement is valid for all organisms, with and without a nervous system.*
>
> — Maturana and Varela (1980)

1.1 Introduction

Following the data-based 'punctuated equilibrium' portrait of evolutionary process uncovered by Eldredge and Gould (1997), abductions of 'big bang' analogs from cosmology have haunted biology, from the 'expanding protein universe' of Dokholyan et al. (2002), through Koonan's (2007) explicit invocation for the major transitions in evolution, studies of 'Cambrian Explosions', and so on (e.g. Koonan et al. 2008; Wallace 2014).

In a somewhat similar way, cognition has become a kind of shibboleth in theoretical biology, seen as the fundamental characterization of the living state at and across its essential scales and levels of organization, as indicated by the quote from Maturana and Varela (1980).

A parallel inference, by Atlan and Cohen (1998) in their study of the immune system, is that cognition, via mechanisms of choice, demands reduction in uncertainty, implying the existence of information sources 'dual' to any cognitive process. A detailed derivation of this inference can be found in (Wallace 2012, Sec. 4).

And as Dretske (1994) indicates, the properties of information sources and transmission in cognitive process are fully constrained by the asymptotic limit theorems of information theory.

Recent developments include the Data Rate Theorem (DRT) determining the minimum rate at which control information must be delivered to stabilize an inherently unstable system (e.g., Nair et al. 2007 and the Mathematical Appendix).

These formalisms provide a scaffold for studying both the dynamics of the living state and the particular characteristics of institutional cognition so critical to all

© The Author(s), under exclusive license to Springer Nature Switzerland AG 2023
R. Wallace, *Essays on the Extended Evolutionary Synthesis*, SpringerBriefs in Evolutionary Biology, https://doi.org/10.1007/978-3-031-29879-0_1

human enterprise. Indeed, the peculiarities of machine cognition do not escape the methodological dragnet (e.g. Wallace 2021a).

With regard to this paper, earlier work (Wallace 2020a, 2021b) outlined how the DRT could be seen as emerging 'naturally' from a fundamental symmetry-breaking closely analogous to the standard argument from physical theory. The symmetries involved, however, were not those most familiar from physical theory—group algebras—but rather the groupoid generalization associated with directed-homotopy equivalence classes of developmental/dynamic paths, in which products are not necessarily defined between all element pairs. The Mathematical Appendix provides an outline of groupoid algebra (Brown 1992; Cayron 2006; Weinstein 1996).

Here, we both extend and simplify that analysis, seeing differentiation between high and low probability paths as the first biological 'big bang'—that of the prebiotic chemical environment—and the imposition of a Data Rate Theorem analog in nonergodic cognitive systems as a second-step 'recombination transparency' transition to structures having a sharp interior/exterior differentiation. The multiple tunable 'global workspaces' of further highly punctuated biological development (e.g. Wallace 2014) then emerge in what can be called a highly natural manner.

The beginning lies in first disjointly separating high and low probability dynamic trajectories, and subsequently separating an inside and outside that necessarily interact. This is not entirely straightforward.

Biology is sometimes seen as haunted by the 'elegance' and 'purity' of whatever is the currently most popular physical theory, leading to reductionist fantasies of deriving biological outcomes from 'elementary first principles', and the author, as a retread physicist, is not entirely immune from such sentiments. Biology and institutional ecology, however, are not particularly subject to reduction. Molecular biology has become an endless swamp and nontrivial abduction of approaches from physical theory to biology—as opposed biophysical applications based heavily in biology itself—are somewhat rare. The 'symmetry breaking' insights we abduct here are consistent with such difficulties, leading to what physical scientists might well view as an impenetrable morass with formal complications that might well challenge cutting-edge string theory.

Some background. As usual, too much for biology, too little for physics.

1.2 Symmetry and Symmetry-Breaking

The Centre for Theoretical Cosmology at Cambridge (CTC 2021) describes the underlying physics case as follows:

> The basic premise of grand unification is that the known symmetries of the elementary particles resulted from a larger (and so far unknown) symmetry group G. Whenever a phase transition occurs, part of this symmetry is lost, so the symmetry group changes. This can be represented mathematically as

$$G \to H \to \ldots \to SU_3 \times SU_2 \times U_1 \to SU_3 \times U_1$$

Here, each arrow represents a symmetry breaking phase transition where matter changes form and the groups—**G, H, SU₃**, etc.—represent the different types of matter, specifically the symmetries that the matter exhibits, and they are associated with the different fundamental forces of nature.

The cosmological significance of symmetry breaking is due to the fact that symmetries are restored at high temperature (just as it is for liquid water when ice melts). For extremely high temperatures in the early Universe, we will even achieve a grand unified state **G**. Viewed from the moment of creation forward, the Universe will pass through a succession of phase transitions at which the strong nuclear force will become differentiated and then the weak nuclear force and electromagnetism...

More generally, Stewart (2017) puts the matter as follows:

Spontaneous symmetry-breaking is a common mechanism for pattern formation in many areas of science. It occurs in a symmetric dynamical system when a solution of the equations has a smaller symmetry group than the equations themselves... This typically happens when a fully symmetric solution becomes unstable and branches with less symmetry bifurcate.

With regard to symmetries and information theory, Yeung (2008) found fundamental relations between information theory inequalities and the theory of finite groups that are similar to those between synchronicity in networks and their permutation symmetries explored by Golubitsky and Stewart (2006). Suppose we have random variables X_1 and X_2 with Shannon uncertainties $H(X_1)$ and $H(X_2)$. The information theory chain rule (Cover and Thomas 2006) finds, for the joint uncertainty $H(X_1, X_2)$, that

$$H(X_1) + H(X_2) \geq H(X_1, X_2) \tag{1.1}$$

This expression has a direct mapping to finite group theory: let G be a finite group, and G_1, G_2 are subgroups of G. Let $\|G\|$ be the order of a group—the number of its elements. Then the intersection $G_1 \cap G_2$ is also a subgroup, and a 'group inequality' can be derived that is the exact analog of Eq. (1.1):

$$\log[\frac{\|G\|}{\|G_1\|}] + \log[\frac{\|G\|}{\|G_2\|}] \geq \log[\frac{\|G\|}{\|G_1 \cap G_2\|}] \tag{1.2}$$

Yeung (2008) defines a probability for a pseudo-random variate associated with a group G as $Pr\{X = a\} = 1/\|G\|$, permitting construction of a group-characterized information source, given that, in general, the joint uncertainty of a set of random variables is not necessarily the logarithm of a rational number. Yeung (2008) shows one-to-one correspondence between unconstrained information inequalities—generalizations of Eq. (1.1)—and finite group inequalities: unconstrained inequalities can be proved by techniques in group theory, and many group-theoretic inequalities can be proven by techniques of information theory. Yeung applies an otherwise obscure unconstrained information inequality to derive, in his Eq. (16.116), a complex and equally obscure—but previously unknown—group inequality.

Extension of the Yeung arguments follows from the fundamental dogma of algebraic topology (Hatcher 2001): that it is possible to form algebraic images of topological spaces. The most basic such image is the fundamental group, leading to Van Kampen's Theorem permitting computation of the fundamental group of spaces that can be decomposed into simpler spaces whose fundamental group is already known. Brown et al. (2011) extend these considerations to groupoids that might be applied to more complicated networks.

Yeung's results suggest information theory-based 'cognitive' generalizations that may extend to essential dynamics of cognition, and to the role of directed homotopy and groupoid symmetry-breaking in those dynamics. Information sources inherently involve directed homotopy via the improbability of 'microreversibility', e.g., in English the sequence of symbols 'the' is far more probable than 'eht'.

Central to any such thinking for the study of cognition, however, is some appropriate definition of 'temperature', and here, for biological (and institutional) purposes, matters become immediately and immensely complicated. We start with a simplified picture of the resource structure in which a real-world cognitive entity must be embedded.

1.3 Resources

At least three resource streams are required by a cognitive entity. The first is measured by the rate $\mathscr{C}$ at which information can be transmitted between elements within the entity, determined as an information channel capacity (Cover and Thomas 2006). The second stream is sensory information regarding the embedding environment, available at a rate $\mathscr{Q}$. The third regards material resources, including metabolic free energy—in a large sense—available at a rate $\mathscr{M}$. Note that *these rates will usually change in time*, leading to the necessity of developing a dynamic theory under highly nonequilibrium circumstances. This is not a trivial matter, and requires some considerable methodological development.

These rates must interact, characterized by a 3 by 3 matrix analogous to, but not the same as, an ordinary correlation matrix. We write this as $\mathbf{Z}$.

An n-dimensional square matrix $\mathbf{M}$ has n scalar invariants $r_1, \ldots, r_n$ defined by the characteristic equation for $\mathbf{M}$:

$$p(\gamma) = \det[\mathbf{M} - \gamma \mathbf{I}] =$$

$$(-1)^n \gamma^n + (-1)^{n-1} r_1 \gamma^{n-1} + (-1)^{n-2} r_2 \gamma^{n-2} - \ldots - r_{n-1}\gamma + r_n \quad (1.3)$$

$\mathbf{I}$ is the n-dimensional identity matrix, det the determinant, and γ a real-valued parameter. The first invariant, r_1, is usually taken as the matrix trace, and the last, r_n, as the determinant.

These scalar invariants make it possible to project the full matrix down onto a single scalar index $M = M(r_1, \ldots, r_n)$ retaining—under some circumstances—much of the basic structure, analogous to conducting a principal component analysis, which does much the same thing for a correlation matrix (e.g. Jolliffe 2002). Wallace (2021a) provides an example in which two such indices are necessary.

With respect to the matrix $\mathbf{Z}$ for the three rates $\mathscr{C}$, $\mathscr{Q}$, and $\mathscr{M}$, the simplest such index might be $Z = \mathscr{C} \times \mathscr{Q} \times \mathscr{M}$. However, there will almost always be important cross-interactions between different resource streams, requiring a more complete analysis, that is, one based on Eq. (1.3) that takes crossterms explicitly into account. That is the assumption we make here.

Clever scalarization—when appropriate—enables approximate reduction to a one-dimensional system. Expansion of Z into vector form is possible, but leads to difficult multidimensional dynamic equations (again, Wallace 2021a).

1.4 Cognition in Nonergodic Systems

Here, we lift the ergodic restriction on information sources (Cover and Thomas 2006). Only in the case that cross-sectional and longitudinal means are the same can information source uncertainty be expressed as a conventional Shannon 'entropy' (Khinchin 1957). We do require that source uncertainties converge for sufficiently long paths, not that they fit some functional form. It is the values of those uncertainties that will be of concern, not their functional expressions. We will study what might be called Adiabatically Piecewise Stationary (APS) systems, in the sense of the Born-Oppenheimer approximation for molecular systems that assume nuclear motions are so slow in comparison with electron dynamics that they can be effectively separated, at least on appropriately chosen trajectory 'pieces' that may characterize the various phase transitions available to such systems. Extension of this work to nonstationary circumstances remains to be done. More specifically, between phase transitions, we assume that the system changes slowly enough so that the asymptotic limit theorems of information and control theories can be invoked.

We carry out this approximation via a fairly standard Morse Function iteration (e.g. Pettini 2007).

Our systems of interest are composed of cognitive submodules that engage in crosstalk. At every scale and level of organization all such submodules are constrained by both their own internals and developmental paths and by the persistent regularities of the embedding environment, including the cognitive intent of adversaries and the regularities of 'grammar' and 'syntax' imposed by embedding evolutionary pressures.

Further, there are structured uncertainties imposed by the large deviations possible within that environment, again including the behaviors of adversaries who may be constrained by quite different developmental trajectories and 'punctuated equilibrium' evolutionary transitions.

Recapitulating somewhat the arguments of Wallace (2018, 2020a), the Morse Function construction assumes a number of interacting factors:

I. As Atlan and Cohen (1998) argue, cognition requires choice that reduces uncertainty. Such reduction in uncertainty directly implies the existence of an information source 'dual' to that cognition at each scale and level of organization. The argument is unambiguous and sufficiently compelling. Again, see (Wallace 2012, Sec. 4) for a more complete explication.

II. Cognitive physiological processes, like the immune and gene expression systems, are highly regulated, in the same sense that 'the stream of consciousness' flows between cultural and social 'riverbanks'. That is, a cognitive information source X_i is generally paired with a regulatory information source X^i.

III. Environments (in a large sense) impose temporal event sequences of very high probability: night follows day, hot seasons follow cold, wet season follows dry, and so on. Thus environments impose their own 'meaningful statements' onto entities and interactions embedded within them via an information source V.

IV. 'Large deviations', following Champagnat et al. (2006) and Dembo and Zeitouni (1998), also involve sets of high probability developmental pathways, often governed by 'entropy'-like laws that imply the existence of yet one more information source L_D.

Full system dynamics must then be characterized by a joint—*path dependent*— nonergodic information source uncertainty

$$H(\{X_i, X^i\}, V, L_D) \tag{1.4}$$

where, here, H is characterized quite generally in terms of the information sources X_i, X^i, V and L_D. Individual dynamic paths can then be assigned a value of that joint source uncertainty, denoted by $H(x)$ for a path x.

This 'fundamental representation' is now defined by individual dynamic path values of source uncertainty and not represented as an 'entropy' function defined for all high-probability paths by an underlying probability distribution (Khinchin 1957). That is, *each path has it's own H-value*, but the functional form of that value is not known in terms of a single underlying probability distribution across all paths.

Again, the set $\{X_i, X^i\}$ includes the internal interactive cognitive dual information sources of the system of interest and their associated regulators, V is taken as the information source of the embedding environment that may include the actions and intents of adversaries/symbionts, as well as 'weather'. L_D is the information source of the associated large deviations possible to the system, possibly including 'punctuated equilibrium' evolutionary transitions.

As above, we project the matrix of essential resources and their interactions onto a scalar rate index Z, according to the argument following Eq. (1.3). This may not always be possible, leading to the multidimensional complexities described in Wallace (2021a).

An essential insight—following from the properties of nonergodic information sources—and a simplification that deviates considerably from the arguments of Wallace (2018), is that the underlying equivalence classes of developmental-behavioral-dynamic system paths used to define groupoid symmetries can now be defined fully in terms of the magnitude of individual path source uncertainties *of individual dynamic paths* $H(x_j)$ such that $x_j = \{x_j^0, x_j^1, \ldots x_j^n, \ldots\}$ at times $m = 0, 1, 2, \ldots n \to \infty$ alone rather than in terms of scalar distortion measures between individual paths as used in the Mathematical Appendix to Wallace (2018). See Khinchin (1957) for details of the nonergodic limit argument. The essential point is that individual paths of sufficient length have associated source uncertainty scalar values that are not, however, calculated as standard Shannon 'entropies' across some probability distribution.

One can envision the equivalence classes of behavioral/developmental paths as defined by the 'game' the organism is playing: growing from inception, foraging for food or habitat, evading predation, wound healing, mating/reproducing, and so on. Paths within each 'game' are taken as equivalent. By the arguments of the Mathematical Appendix, this division of developmental/behavioral paths defines a groupoid. From a human perspective, sets of behavioral paths associated with baseball, football, soccer, rugby, tennis, and so on, are easily discernible and placed in appropriate equivalence classes.

Recall, as well, the conundrum of the ergodic decomposition of nonergodic information sources. It is formally possible to express a nonergodic source as the composition of a sufficient number of ergodic sources, much as it is possible to reduce planetary orbits to a Fourier sum of circular epicycles, obscuring the basic dynamics. Hoyrup (2013) discusses the problem further, finding that ergodic decompositions are not necessarily computable. Here, we finesse the matter by focusing only on the values of the source uncertainties associated with dynamic paths.

1.5 The Prebiotic 'Big Bang'

The next step is to build an *iterated* 'free energy' Morse Function (Pettini 2007) from a Boltzmann pseudoprobability, based on enumeration of high probability developmental pathways x_j, $j = 1, 2, \ldots$ available to the system—each having an individual joint uncertainty $H(x_j) \equiv H_j$ so that

$$P_j = \frac{\exp[-H_j/g(Z)]}{\sum_k \exp[-H_k/g(Z)]} \tag{1.5}$$

where H_j is the source uncertainty of the high probability path j, which—again—we do not assume to be given as a 'Shannon entropy' since we are no longer restricted to ergodic sources.

This step should be recognized as directly imposing a version of the standard 'free energy' in statistical physics as constructed from a partition function (Landau and Lifshitz 2007), using Feynman's (2000) central insight that 'information' can, itself, be viewed as a form of free energy. Hence the iteration.

The essential point at this stage of the argument is a generalization of the fundamental assumption behind the Shannon-McMillan Theorem of information theory Khinchin (1957). That is, in the limit of 'infinite length', *it remains possible to divide the full set of individual dynamic paths into two distinct equivalence classes*, a small set of high probability paths consonant with some underlying 'grammar' and 'syntax' that 'make sense' within the venue of the organism and its environment, and a much larger set of paths of vanishingly low probability not so consonant, essentially a set of measure zero. This is a somewhat subtle point. Observational characterization of such 'grammar' and 'syntax' is not trivial, and in the case of the 'Genetic Code' required much empirical effort (e.g. Marshall 2014).

We thus infer a more general principle.

The temperature-analog characterizing the system, written as $g(Z)$ in Eq. (1.5), can be calculated via a first-order Onsager nonequilibrium thermodynamic approximation built from the partition function, i.e., the denominator of Eq. (1.5) (de Groot and Mazur 1984).

We define the 'iterated free energy' Morse Function F as

$$\exp[-F/g(Z)] \equiv \sum_k \exp[-H_k/g(Z)] \equiv h(g(Z))$$

$$F(Z) = -\log[h(g(Z))]g(Z) \tag{1.6}$$

where the sum is over all possible high probability developmental paths of the system, again, those consistent with an underlying grammar and syntax. To reiterate the central matter—the cognitive 'big bang'—system paths not consonant with grammar and syntax constitute a set of measure zero that has very many more members than the set of high probability paths.

A first, and central, assertion of this analysis is that the possibility of such differentiation—division into at least two equivalence classes—permits emergence from the prebiotic chemical soup of a living state engaging in cognitive behaviors. The differentiation of dynamic paths into high and low probability equivalence classes of behaviors represents the first groupoid structure, analogous to the **G** of Sect. 1.2. This inference, consonant with earlier studies regarding the origin of life, (e.g. Wallace 2011a,b), elevates the asymptotic limit theorems of information theory to a central position in prebiotic studies.

Indeed, Feynman (2000) makes the direct argument that information itself is to be viewed as a form of free energy, using Bennett's clever ideal machine that turns a message directly into work. Here, however, we are concerned with an iterated, rather than a direct, construction.

F, taken as a free energy, then becomes subject to symmetry-breaking transitions as $g(Z)$ varies (Pettini 2007). These symmetry changes, however, are not as associated with physical phase transitions as represented by standard group algebras. Such symmetry changes represent transitions from playing one 'game' to playing another. For example, an organism may engage in foraging behaviors that trigger a predatory attack by another organism. Then the game changes from 'foraging' to 'escape'.

Thus 'cognitive phase changes', as we characterize them here, involve shifts between equivalence classes of high probability developmental/behavioral pathways that are represented as groupoids. To reiterate, this represents a generalization of the group concept such that a product is not necessarily defined for every possible element pair, although multiple products with multiple identity elements are defined (Brown 1992; Cayron 2006; Weinstein 1996). Again, see the Mathematical Appendix for an outline of the theory.

Dynamic equations follow from invoking a first-order Onsager approximation akin to that of nonequilibrium thermodynamics (de Groot and Mazur 1984) in the gradient of an entropy measure constructed from the 'iterated free energy' F of Eq. (1.6). Recall that the central matter of the Onsager approximation is that $\partial Z/\partial t \approx \partial S/\partial Z$. The full algebraic thicket is

$$S(Z) \equiv -F(Z) + Z dF(Z)/dZ$$

$$\partial Z/\partial t \approx dS/dZ = f(Z)$$

$$f(Z) = Z d^2 F/dZ^2$$

$$g(Z) =$$

$$\frac{-C_1 Z - \left(\int \frac{f(Z)}{Z} dZ\right) Z + C_2 + \int f(Z)\, dZ}{RootOf\left(e^Q - h\left(-\frac{C_1 Z + \left(\int \frac{f(Z)}{Z} dZ\right) Z - C_2 - \left(\int f(Z) dZ\right)}{Q}\right)\right)} \tag{1.7}$$

where the last relation follows from an expansion of the third part of Eq. (1.7) using the second expression of Eq. (1.6).

Three important—and somewhat subtle—points:

1. 'It is easily seen that' the 'RootOf' construction generalizes the Lambert W-function (e.g. Yi et al. 2010; Mezo and Keady 2015). Further, since 'RootOf' may have complex number solutions, the temperature analog $g(Z)$ enters the realm of the 'Fisher Zeros' characterizing phase transition in physical systems (e.g. Dolan et al. 2001; Fisher 1965; Ruelle 1964, Sec. 5).
2. Information sources are not microreversible, that is, palindromes are highly improbable, e.g., 'ot' has far lower probability than 'to' in English, so that there are no 'Onsager Reciprocal Relations' in higher dimensional systems. The necessity of groupoid symmetries appears to be driven by this directed homotopy.
3. Characteristically, *there will always be a delay in the rate of provision of Z*, so that, in Eq. (1.7), for example, $f(Z) = \beta - \alpha Z(t)$—an exponential model having $Z(t) = (\beta/\alpha)(1 - \exp[-\alpha t])$—where $Z \to \beta/\alpha$ at a rate determined by α. Other dynamics are possible, such as the 'Arrhenius', $Z(t) = \beta \exp[-\alpha/t]$, with $f(Z) = (Z/\alpha)(\log(Z/\beta))^2$. However Z is constructed from the components $\mathscr{C}$, $\mathscr{D}$ and $\mathscr{M}$, so here, it is the scalar resource rate Z itself that counts.

Suppose, in the first expression of Eq. (1.7), it is possible to approximate the sum across the high probability paths with an integral, so that

$$\exp[-F/g(Z)] \approx \int_0^\infty \exp[-H/g(Z)]dH = g(Z) \qquad (1.8)$$

$g(Z)$ must be real-valued and positive. Then

$$F(Z) = -\log[g(Z)]g(Z)$$

$$g(Z) = -F(Z)/W_L[n, -F(Z)] \qquad (1.9)$$

where W_L is the 'simple' Lambert W-function that satisfies $W_L[n, x] \exp[W_L[n, x]] = x$. It is real-valued only for $n = 0, -1$ and only over limited ranges of x in each case.

In theory, specification of any two of the functions f, g, h permits calculation of the third. h, however, is determined—fixed—by the internal structure of the larger system. Similarly, 'boundary conditions' C_1, C_2 are externally-imposed, further sculpting dynamic properties of the 'temperature' $g(Z)$, and f determines the rate at which the composite essential resource Z can be delivered. Both information and metabolic free energy resources are rate-limited.

1.6   Biological 'Recombination Transparency'

For physical systems, as Sect. 1.2 describes, there will necessarily be a minimum temperature for punctuated activation of the particular set of dynamics associated with a given group structure. For cognitive processes, following the arguments of Eq. (1.7), there will be a minimum necessary value of $g(Z)$ for onset of the next in a series of transitions in the analog to the sequence in Sect. 1.2. That is, at some $T_0 \equiv g(Z_0)$, having a corresponding information source uncertainty H_0, the second groupoid phase transition becomes manifest.

Taking a reaction rate perspective from chemical kinetics (Laidler 1987), we can write an expression for the rate of cognition as

$$L(Z) = \frac{\sum_{H_j > H_0} \exp[-H_j/g(Z)]}{\sum_k \exp[-H_k/g(Z)]} \qquad (1.10)$$

If the sums can be approximated as integrals, then the system's rate of cognition at resource rate index Z can be written as

$$L(Z) \approx \frac{\int_{H_0}^\infty \exp[-H/g(Z)]dH}{\int_0^\infty \exp[-H/g(Z)]dH} = \exp[-H_0/g(Z)]$$

$$= \exp[H_0 W_L(n, -F)/F] \qquad (1.11)$$

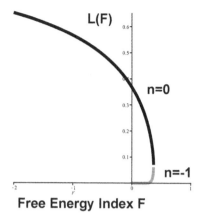

Fig. 1.1 Rate of cognition from Eq. (1.11) as a function of the iterated free energy measure F, taking $H_0 = 1$. The Lambert W-function is only real-valued for orders 0 and -1, and only if $F \leq \exp[-1]$. However, if $F > \exp[-1]$, then a bifurcation instability emerges, with a transition to complex-valued oscillations in cognition rate at higher values. Since F is driven by Z, there is a minimum resource rate for stability

where $W_L(n, -F)$ is the Lambert W-function of order n in the free energy index $F = -\log[g(Z)]g(Z)$, and we enter a brave new world.

Figure 1.1 shows $L(F)$ vs. F, using Lambert W-functions of orders 0 and -1, respectively real-valued only on the intervals $x > -\exp[-1]$ and $-\exp[-1] < x < 0$.

We observe that, here, the Lambert W-function is only real-valued for orders 0 and -1, and only for $F \leq \exp[-1]$. However, if $F > \exp[-1]$, then a bifurcation instability emerges, with a transition to complex-valued oscillations in cognition rate at higher values.

This development recovers what is essentially an analog to the Data Rate Theorem from control theory (Nair et al. 2007, and the Mathematical Appendix), in the sense that the requirement $H > H_0$ in Eqs. (1.10) and (1.11) imposes stability constraints on F, the free energy analog, and by inference, on the resource rate index Z driving it.

1.7 A Simple Application

We can, in fact, parse this basic result further, recovering much of Wallace (2020a), if we again approximate the sum in Eq. (1.6) by an integral—so that $h(g(Z)) = g(Z)$—and make a simple assumption on the form of $dZ/dt = f(Z)$, say $f(Z) = \beta - \alpha Z(t)$. Then

Fig. 1.2 Classic signal transduction for the cognition rate from Eq. (1.12), setting $\alpha = 1$, $C_1 = -2$, $C_2 = 2$, $H_0 = 1$. A changing β is taken as the 'arousal' measure. Boundary conditions are appropriate to a signal transduction model

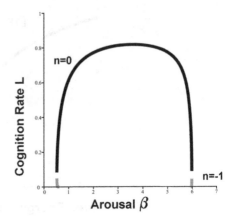

$$g(Z) = -\frac{2\ln(Z)\,Z\beta - Z^2\alpha + 2C_1Z - 2Z\beta + 2C_2}{2W_L\left(n, -\ln(Z)\,Z\beta + \frac{Z^2\alpha}{2} - C_1Z + Z\beta - C_2\right)} \qquad (1.12)$$

with, again, $L = \exp[-H_0/g(Z)]$, depending on the rate parameters α and β, the boundary conditions C_i, and the degree of the Lambert W-function. Proper choice of boundary conditions generates a classic signal transduction analog (e.g. Wallace 2020a, 2021c). That is, since $Z \to \beta/\alpha$, we look at the cognition rate for fixed α and boundary conditions C_j as β increases. The result is shown in Fig. 1.2, for appropriate boundary conditions.

Similar results follow if $\exp[-F/g(Z)] = h(g(Z)) \propto A_m g(Z)^m$, that is, if the function $h(g(Z))$ has a strongly dominant term of order $m > 0$.

1.8   Specialization and Cooperation: Multiple Workspaces

The emergence of a generalized Lambert W-function in Eq. (1.7), reducing to a 'simple' W-function if $h(g(Z)) = g(Z)$, is suggestive. Recall that the fraction of network nodes $\mathcal{N}$ in a giant component of a random network of N nodes with probability P of linkage between them can be given as (Newman 2010)

$$\mathcal{N} = \frac{1}{NP}\left(W(0, -NP\exp[-NP]) + NP\right)$$

$$P = \frac{\log(-\frac{1}{\mathcal{N}-1})}{\mathcal{N}N} \propto \frac{1}{N} \qquad (1.13)$$

where, again, the Lambert W-function emerges.

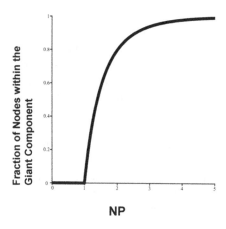

Fig. 1.3 Fraction of an N-node random network within the giant component as determined by the probability of contact P between nodes. The essential point is the punctuated accession to 'global broadcast' if and only if $NP > 1$ (e.g. Baars 1989; Dehaene and Changeux 2011). We might expect P to be a monotonic increasing function of $g(Z)$. Compare the second form of Eq. (1.13) with models of percolation on rough 'fitness landscapes' (e.g. Gavrilets 2010)

This expression has punctuated onset of a giant component of linked nodes only for $NP > 1$. See Fig. 1.3. In general, we might expect P to be a monotonic increasing function of $g(Z)$.

Note that, for any given fraction $\mathcal{N}$, the needed probability threshold scales as $1/N$, consonant with models of percolation on rough high dimensional 'fitness landscapes' (e.g. Gavrilets 2010, and references therein).

The line of argument is that, within organisms (or institutions), interacting cognitive submodules can become linked into shifting, tunable, temporary, workgroup equivalence classes to address similarly rapidly shifting patterns of threat and opportunity. These might range from complicated but relatively slow multiple global workspace processes of gene expression and immune function to the rapid—hence necessarily stripped-down—single-workspace neural phenomena of higher animal consciousness (Wallace 2012).

More complicated approaches to such a phase transition—involving Kadanoff renormalizations of the Morse Function free energy measure F and related measures—can be found in Wallace (2005, 2012, 2017, 2022). The essential idea is to adapt Wilson's (1971) standard treatment of phase transition in physical systems to information systems, invoking Feynman's (2000) identification of information as another kind of free energy.

By contrast here, while multiple workspaces are most simply invoked in terms of a simultaneous set of the $g_j(Z_j)$, where the j refers to a particular workspace—e.g., envision an immune system engaging in routine cellular maintenance in one location while fighting off a pathogen in some other—individual workspace tunability emerges from exploring equivalence classes of network topologies associated with a particular value of some designated $g_j(Z_j)$. Central matters then revolve around the equivalence class decompositions implied by the existence of the resulting workgroups, leading again to dynamic groupoid symmetry-breaking as they shift form and function in response to changing patterns of threat and opportunity. See Wallace (2021b, Sec. 12) for a parallel argument from an ergodic system perspective. Here, the $g(Z)$ substitute for the 'renormalization constant' ω in that development. For brevity, we omit a full discussion here, but the essential idea is that, in the original Wilson/Kadanoff model (Wilson 1971), ω is the 'ordinary' physical temperature. Here, $g(Z)$ serves that function..

We can, indeed, calculate the cognition rate of a single 'global workspace' as follows.

Suppose we have N linked cognitive submodules in a giant component, i.e., we operate in the upper portion of Fig. 1.3. The free energy Morse Function F can be expressed in terms of a full-bore partition function as

$$\exp[-F/g(Z)] = \sum_{k=1}^{N} \sum_{j=1}^{M} \exp[-H_{k,j}/g(Z)]$$

$$\approx \sum_{k=1}^{N} \int_0^\infty \exp[-H_k/g(Z)]dH_k = \sum_{k=1}^{N} g(Z) = Ng(Z)$$

$$F = -\log[Ng(Z)]g(Z), \quad g(Z) = \frac{-F}{W_L(n, -NF)} \tag{1.14}$$

The sum over j represents available states within individual submodules, and the sum over k is across submodules, and W_L is again the Lambert W-function. Note, however, the appearance of the factor NF in the last expression.

Taking H_0^k as the 'activation energy' for the 'reaction rate' cognition model, the rate of cognition of the linked-node giant component can be expressed as

$$L = \frac{\sum_{k=1}^{N} \int_{H_0^k}^\infty \exp[-H_k/g(Z)]dH_k}{\sum_{k=1}^{N} \int_0^\infty \exp[-H_k/g(Z)]dH_k}$$

$$= \frac{g(Z) \sum_k \exp[-H_0^k/g(Z)]}{Ng(Z)}$$

$$= \left(\sum_k \exp[-H_0^k/g(Z)] \right)/N \equiv < L_k > \tag{1.15}$$

where, not entirely unexpectedly, $< L_k >$ represents an averaging operation. More sophisticated averages, of course, could be applied, at the expense of considerably more formalism.

If we impose the approximation of Onsager nonequilibrium thermodynamics, i.e., defining $S(Z) = -F(Z) + ZdF/dZ$, and assume $dS/dZ = f(Z)$, we again obtain $f(Z) = Z d^2 F/dZ^2$, and can calculate $g(Z)$ from the last expression in Eq. (1.14).

The appearance of N in expressions for $g(Z)$ and L is of some note. In particular, groupthink—high values of H_0^k—may result in failure to detect important signals.

Multiple workspaces present a singular problem of resource delivery.

If we assume an overall limit to available resources across a multiple workspace system $j = 1, 2, \ldots$, so that $Z = \sum_j Z_j$, then it is possible to carry out a simple Lagrangian optimization on the rate of system cognition $\sim \sum_j \exp[-H_j^0/g_j(Z_j)]$ as

$$\mathcal{L} \equiv \sum_j \exp[-H_j^0/g_j(Z_j)]) + \lambda \left(Z - \sum_j Z_j \right)$$
$$\partial \mathcal{L}/\partial Z_j = 0 \qquad (1.16)$$

leading, after some development, to a necessary expression for individual subsystem cognition rates as

$$\exp[-H_j^0/g_j(Z_j)] = \lambda \left(Z_j + C_j \right) \geq 0 \qquad (1.17)$$

The C_j are appropriate boundary condition and λ is to be viewed in economic terms as the *shadow price* imposed by 'environmental constraints', in a large sense (e.g., Jin et al. 2008; Robinson 1993).

The essential point is that cognitive dynamics—the rate of cognition—in this model, is strongly driven by λ, to be interpreted as an environmental signal. Depending on the boundary conditions, if the environmental signal is not of the correct sign and value, no cognitive detection will take place.

1.9 Discussion

We are, then, led toward a generalization of the physical symmetry-breaking progression in Sect. 1.2, but in terms of a series of groupoid/equivalence class transitions associated with biological and other forms of cognition:

High vs. Low probability paths → Interior and Exterior Interact → Multiple Interacting Tunable Workspaces

The first groupoid phase transition—the 'big bang' of the primordial prebiotic chemical environment—instantiates a generalized form of the Shannon-McMillan Source Coding Theorem: 'grammar' and 'syntax' emerge in sequential chemical processes. 'Ungrammatical statements' become a set of measure zero.

The second transition, differentiating interior from exterior while permitting interaction between them, imposes a version (or versions) of the Data Rate Theorem: inner/outer interaction implies the necessity of control under uncertainty. Indeed, cognition and its regulation become closely coupled: ultimately, the 'stream of consciousness' always has social and cultural riverbanks confining its meanderings.

The third transition represents onset of the trajectory to gene expression, multicellularity, organ distinction, speciation, and, in higher animals, consciousness. However, consciousness, through its 100ms time constant, is a necessarily stripped-down and greatly simplified version of the slower, but far more complicated, multi-workspace phenomena (Wallace 2012, 2022).

Social, institutional, and machine parallels seem obvious (e.g. Wallace 2015, 2020b, 2021a).

However, biological (and institutional) processes tend to be highly complex and very subtle, and analogies with current trends in physical theory are, at best, strained. Thus, it seems unlikely that groupoid symmetry breaking is the end of the matter. Indeed, more general symmetries can be expected to emerge from further study, particularly for information sources that are not stationary. These will, perhaps, be based on semigroupoids, or even less specified, algebraic structures. This work remains to be done.

In any event, the classes of probability models explored here might well be developed into statistical tools for the analysis of real-time and other data across a spectrum of important disciplines confronted by cognition and its manifold dysfunctions.

1.10   Mathematical Appendix

Groupoids

We follow Brown (1992) closely. Consider a directed line segment in one component, written as the source on the left and the target on the right.

$$\bullet \longrightarrow \bullet$$

Two such arrows can be composed to give a product **ab** if and only if the target of **a** is the same as the source of **b**

Brown puts it this way,

One imposes the geometrically obvious notions of associativity, left and right identities, and inverses. Thus a groupoid is often thought of as a group with many identities, and the reason why this is possible is that the product **ab** is not always defined.

We now know that this apparently anodyne relaxation of the rules has profound consequences...[since] the algebraic structure of product is here linked to a geometric structure, namely that of arrows with source and target, which mathematicians call a directed graph.

Cayron (2006) elaborates this as follows,

A group defines a structure of actions without explicitly presenting the objects on which these actions are applied. Indeed, the actions of the group G applied to the identity element e implicitly define the objects of the set G by ge = g; in other terms, in a group, actions and objects are two isomorphic entities. A groupoid enlarges the notion of group by explicitly introducing, in addition to the actions, the objects on which the actions are applied. By this approach, many identities may exist (they correspond to the actions that leave an object invariant).

It is of particular importance that equivalence class decompositions permit construction of groupoids in a highly natural manner.

Weinstein (1996) and Golubitsky and Stewart (2006) provide more details on groupoids and on the relation between groupoids and bifurcations.

An essential point is that, since there are no necessary products between groupoid elements, 'orbits', in the usual sense, disjointly partition groupoids into 'transitive' subcomponents.

The Data Rate Theorem

Real-world environments are inherently unstable. Organisms, to survive, must exert a considerable measure of control over them. These control efforts range from immediate responses to changing patterns of threat and affordance, through niche construction, and, in higher animals, elaborate, highly persistent, social and sociocultural structures. Such necessity of control can, in some measure, be represented by a powerful asymptotic limit theorem of probability theory different from, but as fundamental as, the Central Limit Theorem: the Data Rate Theorem, first derived as an extension of the Bode Integral Theorem of signal theory.

Consider a reduced model of a control system as follows:

For the inherently unstable system of Fig. 1.4, assume an initial n-dimensional vector of system parameters at time t, as x_t. The system state at time $t + 1$ is then—

Fig. 1.4 The reduced model of an inherently unstable system stabilized by a control signal U_t (Wallace 2020c, fig. 4)

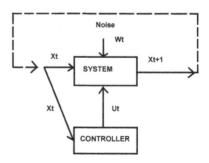

near a presumed nonequilibrium steady state—determined by the first-order relation

$$x_{t+1} = \mathbf{A}x_t + \mathbf{B}u_t + W_t \tag{1.18}$$

In this approximation, $\mathbf{A}$ and $\mathbf{B}$ are taken as fixed n-dimensional square matrices. u_t is a vector of control information, and W_t is an n-dimensional vector of Brownian white noise.

According to the DRT, if H is a rate of control information sufficient to stabilize an inherently unstable control system, then it must be greater than a minimum, H_0,

$$H > H_0 \equiv \log[\| \det[\mathbf{A}^m] \|] \tag{1.19}$$

where det is the determinant of the largest subcomponent $\mathbf{A}^m$—with $m \leq n$—of the matrix $\mathbf{A}$ that has eigenvalues ≥ 1. H_0 is defined as the rate at which the unstable system generates 'topological information' on its own. See Nair et al. (2007) for details of the calculation.

If this inequality is violated, stability fails.

References

Atlan H., and I. Cohen. 1998. Immune information, self-organization, and meaning. International Immunology 10:711–717.

Baars B. 1989. *A Cognitive Theory of Consciousness*. New York: Cambridge University Press.

Brown, R. 1992. Out of line. Royal Institute Proceedings 64:207–243.

Brown, R., P. Higgins, and R. Sivera. 2011. *Nonabelian Algebraic Topology: Filtered Spaces, Crossed Complexes, Cubical Homotopy Groupoids*. EMS tracts in mathematics, Vol. 15.

Cayron, C. 2006. Groupoid of orientational variants. Acta Crystalographica Section A A62:21040.

Champagnat, N., R. Ferriere, and S. Meleard. 2006. Unifying evolutionary dynamics: from individual stochastic process to macroscopic models. Theoretical Population Biology 69:297–321.

Cover, T., and J. Thomas. 2006. *Elements of Information Theory*, 2nd ed. New York: Wiley.

CTC. 2021. http://www.ctc.cam.ac.uk/outreach/origins/cosmic_structures_one.php

de Groot, S., and P. Mazur. 1984. *Nonequilibrium Thermodynamics*. New York: Dover.

Dehaene, S., and J. Changeux. 2011. Experimental and theoretical approaches to conscious processing. Neuron 70:200–227.

Dembo, A., and O. Zeitouni. 1998, *Large Deviations and Applications*, 2nd ed. New York: Springer.

Dokholyan, N., B. Shakhnovich, and E. Shakhnovich. 2002. Expanding protein universe and its origin from the biological Big Bang. Proceedings of the National Academy of Sciences 99:14132–14136.

Dolan, B., W. Janke, D. Johnston, and M. Stathakopoulos. 2001. Thin Fisher zeros. Journal of Physics A 34:6211–6223.

Dretske, F. 1994. The explanatory role of information. Philosophical Transactions of the Royal Society A 349:59–70.

Eldredge, N., and S. Gould. 1997. On punctuated equilibria. Science 276:338–341.

Feynman, R. 2000. *Lectures in Computation*. Boulder: Westview Press.

Fisher, M. 1965. *Lectures in Theoretical Physics*, Vol. 7. Boulder: University of Colorado Press.

Gavrilets, S. 2010. High-dimensional fitness landscapes and speciation. In *Evolution: The Extended Synthesis*, ed. Massimo Pigliucci and Gerd B. Müller. Cambridge : MIT Press.

Golubitsky, M., and I. Stewart. 2006. Nonlinear dynamics and networks: the groupoid formalism. Bulletin of the American Mathematical Society 43:305–364.

Hatcher, A. 2001. *Algebraic Topology*. New York: Cambridge University Press.

Hoyrup, M. 2013. Computability of the ergodic decomposition. Annals of Pure and Applied Logic 164:542–549.

Jin, H., Z. Hu, and X. Zhou. 2008. A convex stochastic optimization problem arising from portfolio selection. Mathematical Finance 18:171–183.

Jolliffe, I. 2002. *Principal Component Analysis*. New York: Springer.

Khinchin, A. 1957. *Mathematical Foundations of Information Theory*. New York: Dover.

Koonan, E. 2007. The Biological Big Bang model for the major transitions in evolution. Biology Direct. http://www.biology-direct.com/content/2/1/21

Koonan, E., Y. Wolf, K. Nagasaki, and V. Dolja. 2008. The Big Bang of picorna-like virus evolution antedates the radiation of eukaryotic supergroups. Nature Reviews: Microbiology 6:925–939.

Laidler, K. 1987. *Chemical Kinetics*, 3rd ed. New York: Harper and Row.

Landau, L., and E. Lifshitz. 2007. *Statistical Physics*, 3rd ed., Part 1. New York: Elsevier.

Marshall, J. 2014. The genetic code, PNAS 111:5760.

Maturana, H., and F. Varela. 1980. *Autopoiesis and Cognition: The Realization Of The Living*. Boston: Reidel.

Mezo I., and G. Keady. 2015. Some physical applications of generalized Lambert functions. arXiv:1505.01555v2 [math.CA] 22 Jun 2015.

Nair, G., F. Fagnani, S. Zampieri, and R. Evans. 2007. Feedback control under data rate constraints: an overview. Proceedings of the IEEE 95:108137.

Newman, M. 2010. *Networks: An Introduction*. New York: Oxford University Press.

Pettini, M. 2007. *Geometry and Topology in Hamiltonian Dynamics and Statistical Mechanics*. New York: Springer.

Robinson, S. 1993. Shadow prices for measures of effectiveness II: general model. Operations Research 41:536–548.

Ruelle, D. 1964. Cluster property of the correlation functions of classical gases. Reviews of Modern Physics 36:580–584.

Stewart, I. 2017. Spontaneous symmetry-breaking in a network model for quadruped locomotion. International Journal of Bifurcation and Chaos 14:1730049.

Wallace, R. 2005. *Consciousness: A Mathematical Treatment of the Global Neuronal Workspace Model*. New York: Springer.

Wallace, R. 2011a. On the evolution of homochirality. Comptes Rendus Biologies 334:263–268.

Wallace, R., 2011b. Structure and dynamics of the 'protein folding code' inferred using Tlusty's topological rate distortion approach. BioSystems 103:18–26.

Wallace, R. 2012. Consciousness, crosstalk, and the mereological fallacy: an evolutionary perspective. Physics of Life Reviews 9:426–453.

Wallace, R. 2014. A new formal perspective on 'Cambrian Explosions'. Comptes Rendus Biologies 337:1–5.

Wallace, R. 2015. *An Ecosystem Approach to Economic Stabilization: Escaping the Neoliberal Wilderness*. New York: Routledge.

Wallace, R. 2017. *Computational Psychiatry: A Systems Biology Approach to the Epigenetics of Mental Disorders*. New York: Springer.

Wallace, R. 2018. New statistical models of nonergodic cognitive systems and their pathologies. Journal of Theoretical Biology 436:72–78.

Wallace, R. 2020a. On the variety of cognitive temperatures and their symmetry-breaking dynamics. Acta Biotheoretica. https://doi.org/10.1007/s10441-019-09375-7

Wallace, R. 2020b. *Cognitive Dynamics on Clausewitz Landscapes: The Control and Directed Evolution of Organized Conflict*. New York: Springer.

Wallace, R. 2020c. Signal transduction in cognitive systems: origin and dynamics of the inverted-U/U dose-response relation. Journal of Theoretical Biology 504:110377.

Wallace, R. 2021a. How AI founders on adversarial landscapes of fog and friction. *Journal of Defense Modeling and Simulation* 19. https://doi.org/10.1177/1548512920962227

Wallace, R. 2021b. Toward a formal theory of embodied cognition. *BioSystems* 202:104356.

Wallace, R. 2021c. Embodied cognition and its pathologies: the dynamics of institutional failure on wickedly hard problems. *Communications in Nonlinear Science and Numerical Simulation* 95:105616.

Wallace, R. 2022. *Consciousness, Cognition, and Crosstalk: The Evolutionary Exaptation of Nonergodic Groupoid Symmetry-Breaking*. New York: Springer.

Wilson K. 1971. Renormalization group and critical phenomena. I Renormalization group and the Kadanoff scaling picture. *Physics Reviews B* 4:3174–3183.

Weinstein, A. 1996. Groupoids: unifying internal and external symmetry. *Notices of the American Mathematical Association* 43:744–752.

Yeung, H. 2008. *Information Theory and Network Coding*. New York: Springer.

Yi, S., P.W. Nelson, and A.G. Ulsoy. 2010. *Time-Delay Systems: Analysis and Control Using the Lambert W Function*. New Jersey: World Scientific.

Chapter 2
On the Extended Evolutionary Synthesis

The process of evolution is a complex set of phenomena posing a diverse array of questions that requires dissection from many angles simultaneously... [T]here is no 'fundamental' viewpoint or level to which we can reduce our picture... [A] fully unified view of evolutionary processes may be out of reach...

— Love (2010)

2.1 Introduction

Laland et al. (2014), in a famous broadside, frame the central problem of contemporary evolutionary theory in biology this way,

> ...[M]ainstream evolutionary theory has come to focus almost exclusively on genetic inheritance and processes that change gene frequencies...[A] 'gene-centric' focus fails to capture the full gamut of processes that direct evolution. Missing pieces include how physical development influences the generation of variation (developmental bias); how the environment directly shapes organisms; traits (plasticity); how organisms modify environments (niche construction); and how organisms transmit more than genes across generations (extra-genetic inheritance)...

More detailed such analyses abound (e.g. Laland et al. 1999, 2015; Muller 2017; Odling-Smee et al. 2003; Pigliucci and Muller 2010), leading to emergence of an Extended Evolutionary Synthesis (EES) in biology that focuses as much on development and environment as on genetic heritage.

A perceived limitation of this version of an EES has been the apparent lack of a sufficiently elegant and comprehensive mathematical formulation matching that gracing (or, as many argue, fatally burdening) the innumerable gene frequency treatments of evolutionary dynamics. What might be criticized as piecemeal treatments do indeed abound (e.g., EES 2020; Pigliucci and Muller 2010), but the sense of 'something big' still seems missing, in comparison with the gene frequency-based analysis.

It can, however, be seriously argued that a single unitary and fully comprehensive 'something big' may simply not exist for the almost infinite varieties of evolutionary process active at and across the full range of scales of biological, social, institutional, and cultural phenomena. The inglorious demise of 'meme theory' should have raised a red flag that narrow gene-centered models of evolutionary process may indeed have serious limitations (Chvaja 2020).

While 'Big' monolithic conceptions may be fundamentally inappropriate for any form of ESS, we argue here that it may be possible to construct something far more modest that encompasses enough of the underlying ideas to be of intellectual interest, at least from the perspective of data analysis. That is, here we will forge the asymptotic limit theorems of information (and control) theory into new statistical tools that might find use uncovering pattern and process across a broader range of evolutionary phenomena than can be encompassed by gene frequency models alone.

A number of recently-published materials—(R. Wallace and R.G. Wallace 1998; 1999, R. Wallace 2002, 2009, 2010, 2011a,b,c, 2012a,b, 2013a,b,c, 2016a, R. Wallace and D. Wallace 1998; 2008; 2009a; 2011; 2016, R. Wallace, D. Wallace, and R.G. Wallace 2009, Glazebrook and Wallace 2012, and R.G. Wallace and R. Wallace 2009b)—outline a possible approach. The essence lies in first recognizing that the living state is cognitive, in a formal sense, at all scales and levels of organization (e.g. Maturana and Varela 1980), and in the characterization of such cognition through 'dual' information sources constrained by the asymptotic limit theorems of information theory (Dretske 1994; Atlan and Cohen 1998; Wallace 2012a,b).

What follows is, at least mathematically, surprisingly straightforward: embedding ecosystems, niches, genetic, chemical epigenetic, and cultural, heritage transmission, and the 'large deviations' that afflict them all, can be viewed in terms of information sources. Interaction occurs through an appropriate joint information source, and dynamics emerge via abduction of approximate methods from statistical mechanics and nonequilibrium thermodynamics. The lack of microreversibility inherent to information transmission—'directed homotopy'—leads to groupoid-based symmetry-breaking phase transition analogs more familiar as 'punctuated equilibrium'.

2.2 First Notions

General Information Sources

The essential character of an information source (Khinchin 1957; Cover and Thomas 2006) is the ability to divide streams of output into two sets, a small one of high probability, containing a relatively few 'meaningful' statements consonant with an underlying grammar and syntax (in a large sense), and a much larger 'nonsense' set of vanishingly small probability. This is the fundamental content of the Shannon-McMillan Theorem.

That is, underlying grammar and syntax are inherent to meaningful sequences associated with an information source.

For a stationary, 'ergodic' system in which cross-sectional probabilities match longitudinal probabilities, it is further possible to define a system-wide source uncertainty for an information source X as

$$H[X] = \lim_{n \to \infty} \frac{\log[N(n)]}{n} \tag{2.1}$$

where $N(n)$ is the number of 'meaningful sequences' of length n. This value is the same across all possible output sequences $x^n = \{x_0, x_1, \ldots x_n \ldots\}$, $n \to \infty$ and can be expressed in terms of 'Shannon entropies' having the form $-\sum_j P_j \log(P_j)$, where the P_j constitute an appropriate probability distribution.

For a stationary, non-ergodic system, however, each limiting path x^n, as $n \to \infty$ will have its own path-dependent H-value (Khinchin 1957), not expressible as a Shannon entropy across a probability distribution.

While it is, in theory, possible to represent any non-ergodic information source as an appropriate sum or integral of ergodic sources, in our context, this would be like making a Ptolemaic epicycle expansion of Keplerian orbits. Here, via the nonequilibrium thermodynamics formalism, we will impose something of a Keplerian simplicity on non-ergodic information sources.

Biological Information Sources

1. Natural ecosystem observations indeed have grammar and syntax. Night follows day. Much off the Equator, seasonal variations follow predictable patterns regarding insolation, temperature, rain/snow, and the like. Within a limited ecosystem, on a limited time scale, there are expected flora and fauna that interact (roughly) in familiar patterns—mutualism, predation, symbiosis, and so forth. Call this information source Y. As an extension of theory, Y itself might well be broken into interacting niche substructure information sources.

2. Genetic heritage can be treated as information transmission. Adami et al. (2000), Ofria et al. (2003), Adami and Cerf (2000) provide detailed arguments. Indeed, the transmission of genetic information is much a contextual matter involving an information source that must interact with a broad spectrum of embedding ecosystem information sources. Wallace and Wallace (2009b, Ch.1) provide a fuller exposition.

3. Gene expression is itself a cognitive phenomenon involving a dual information source. As Wallace (2010) puts it, a cognitive paradigm is needed to understand gene expression, much as Atlan and Cohen (1998) invoke a cognitive paradigm for the immune system. Cohen and Harel (2007) assert that gene expression is a reactive system that calls our attention to its emergent properties, i.e., behaviors that, taken as a whole, are not expressed by any one of the lower scale components that comprise it. Cellular processes react to both internal and external signals to produce diverse tissues internally, and diverse general phenotypes across various scales of space, time, and population, all from a single set or relatively narrow distribution of genes.

4. The essence of the Atlan/Cohen cognitive paradigm is that cognition involves choice of a smaller set of actions from a much larger set available to a cognitive entity. Choice reduces uncertainty, and reduction in uncertainty implies the

existence of an information source 'dual' to the cognitive process. The argument is direct, compelling, and intuitive.

5. Cognitive processes on inherently unstable dynamic 'roadways' are almost always paired with regulatory information sources: blood pressure must not become excessive, immune systems must not attack self-tissues, consciousness in higher animals must often be paired with internalized social control, cancerous cells must be constrained, and so on. If X_i is an information source within an organism, we can expect a parallel regulatory source as X^i.

6. Large deviations (Champagnat et al. 2006; Dembo and Zeitouni 1998) follow high probability developmental pathways governed by entropy-like laws that imply the existence of another information source, say L_D.

7. For individual and associated collections of species, 'selection pressures' are not always random, but are usually highly structured so that sequences of evolutionary challenge can be associated with an information source $\mathscr{S}_P$, again allowing identification 'meaningful' sequences according to underlying grammar and syntax.

Coevolutionary Stochastic Burden

Dieckmann and Law (1996) argue at some length that the study of what they call 'asymptotic stationary states' or 'fixed points'—quasi-stable conditions in evolutionary process—requires address of four critical matters:

1. The evolutionary process needs to be considered in a coevolutionary context.
2. A proper mathematical theory of evolution should be dynamical.
3. The coevolutionary dynamics ought to be underpinned by a microscopic theory.
4. The evolutionary process has important stochastic elements.

We will, in fact, recover something like the Dieckmann and Law (1996) perspective in Eq. (2.9) below, but without using the canonical equation of evolutionary game dynamics.

In what follows, then, we attempt to span these two sets of points in a comprehensive 'mathematically elegant' manner. Other approaches, of course, might well be taken e.g., Gonzalez-Forero and Gardner (2021, 2022), who extend evolutionary game dynamics results.

2.3 The Basic Theory

Ecosystem-embedded coevolutionary process, as we have characterized it here, can thus be described by a joint information source having an uncertainty (Cover and Thomas 2006):

$$H(\{X_i, \ X^i\}, Y, L_D, \mathscr{S}_P) \tag{2.2}$$

that may, under nonergodic circumstance, be path-by-path.

Indeed, such a source is unlikely to be ergodic and thus cannot be characterized as Shannon entropies across some appropriate probability distribution. That is, each joint path converges to an individual source uncertainty value as sequence length increases. See Khinchin (1957) for a detailed discussion. For nonergodic information sources it remains possible to assign a path dependent information source uncertainty to each possible—sufficiently long—high-probability path.

The set $\{X_i, X^i\}$ is taken to pair the basic cognitive biological process X_i with a regulatory process X^i: for biological systems, major cognitive phenomena are almost always paired with essential regulators, as in gene expression and immune function across the life course. It can be argued that progressive failure of such bioregulators in higher animals most often drives diseases of aging.

We now construct an index of the rate at which essential resources can be delivered to elements within the coevolutionary ecosystem of interest. This involves scalarization of a possible multidimensional set of resources including the basic metabolic free energy sources, ecosystem richness in terms of predation, mutualism, symbiosis, and the like.

In defining this index, we seek a scalar measure built from n critical rate parameters. There will be, then, an $n \times n$ matrix $\mathbf{Z}$ of 'main' and 'interaction' effects analogous to—but perhaps much different from—a correlation matrix. That is, the matrix $\mathbf{Z}$ is not likely to be symmetric.

It is, however, possible to determine n matrix invariants r_i according to the standard polynomial construction

$$p(\gamma) = \det(\mathbf{Z} - \gamma \mathbf{I}) = \gamma^n + r_1 \gamma^{n-1} + \ldots + r_{n-1}\gamma + r_n \quad (2.3)$$

where n is the order of the matrix, det is the determinant, and $\mathbf{I}$ the $n \times n$ identity matrix. The first invariant is the $\mathbf{Z}$-matrix trace, and the last is $\pm$ the determinant. Based on the n invariants, it becomes possible to construct an appropriate scalar index $Z = Z(r_1, \ldots, r_n)$. One analogy is the magnitude of the largest component in a standard Principal Component Analysis.

Matters can become extortionately complicated. See, for example, computational models of mitochondrial function by Wu et al. (2007), who consider 64 state variables and 210 parameters, and by Bazil et al. (2020), considering 73 and 359, respectively. The scalar parameter Z thus encompasses considerable hidden machinery whose unpacking may be central to developing useful statistical models. In particular, it seems likely necessary to extend the formalisms presented here to analogs of Z that are more complicated algebraic objects than scalars. Following Wallace (2021a), we will reconsider something of this in more detail at a later stage of the argument.

Feynman (2000), building on work by Bennett, shows how to construct an ideal machine that converts information into useful work, i.e., free energy. This, it turns out, is a central insight: information can be seen as a form of free energy that we now compound, building an iterated free energy Morse Function (Pettini 2007) characterizing a particular coevolutionary ecosystem. This is done using a Boltzmann probability expression.

The first step is to enumerate the high probability developmental pathways available to the system. Taking $j = 1, 2, \ldots$, it is possible to define a probability P_j for a particular path j as

$$P_j = \frac{\exp[-H_j/g(Z)]}{\sum_k \exp[-H_k/g(Z)]} \equiv \frac{\exp[-H_j/g(Z)]}{\mathscr{H}(g(Z))} \tag{2.4}$$

where $\mathscr{H}(g(Z))$ is this development's analog to the usual statistical mechanical partition function (Landau and Lifshitz 1980).

The analysis applies to nonergodic as well as to ergodic information sources and can be used for systems in which each developmental pathway x_j has its own source uncertainty measure H_{x_j}. Again, this value can only be defined as a 'Shannon entropy' for an ergodic system.

The 'temperature' analog $g(Z)$ in the expression has to be calculated from first principles. We will do so by imposing Onsager-like system dynamics built from the partition function.

A classic Morse Function (Pettini 2007) 'iterated free energy' F can be written in terms of the partition function denominator of Eq. (2.4) as

$$\exp[-F/g(Z)] \equiv \sum_k \exp[-H_k/g(Z)] = \mathscr{H}(g(Z))$$

$$F = -\log[\mathscr{H}(g(Z))]g(Z)$$

$$g(Z) = \frac{-F}{Root Of\,(\exp[Y] - \mathscr{H}(-F/Y))}$$

$$\tag{2.5}$$

If $\mathscr{H}(g(Z)) = g(Z)$ then $g(Z) = -F/W(n, -F)$, where $W(n, x)$ is the Lambert W-function of order n that satisfies the relation $W(n, x) \exp[W(n, x)] = x$. It is real-valued only for $n = 0, -1$ respectively over the ranges $x \geq -\exp[-1]$ and $-\exp[-1] \leq x \leq 0$.

The RootOf construction is thus to be interpreted as a generalized Lambert W-function.

This result is deceptively simple, having a number of deep and subtle implications:

1. The RootOf construction may have complex number solutions so that the temperature-analog function $g(Z)$ must be taken as analogous to the highly counterintuitive Fisher Zeros that characterize phase transitions in physical systems (Dolan et al. 2001; Fisher 1965; Ruelle 1964).
2. Since information sources are fundamentally dissipative—palindromes are vanishingly rare and directed homotopy dominates—microreversibility is impossible. As a direct consequence, there can be no 'Onsager Reciprocal Relations' in systems of dimension greater than one.

3. F, as a Morse Function, is subject to symmetry-breaking transitions as $g(Z)$
   varies (Pettini 2007). The symmetries here, however, are not those of physical
   phase transitions most often represented by standard group structures. Information system phase change involves punctuated transitions between equivalence
   classes of high probability, path-dependent and directed signal sequences, necessarily represented as groupoids. Groupoids are extensions of 'ordinary' algebraic
   groups in which a product is not necessarily defined for every possible pair
   of elements (Brown 1992; Cayron 2006; Weinstein 1996). The emergence of
   groupoids appears to be a consequence of the inherently one-way directed
   homotopy of information sources.

That is, for coevolutionary phenomena, groupoid symmetries are inherent to the
directed homotopy-induced by failure of local time reversibility for information
systems. This occurs because palindromes have vanishingly small probability. For
example, in English, 'the' has meaning in context while 'eht' has vanishingly low
probability in real-world communication.

More complicated information systems may even require more general structures, such as small categories or semigroupoids, further complicating analogs to
the standard symmetry-breaking dynamics of physical systems.

Thus there may be a number of phase analogs available to a coevolutionary
system as $g(Z)$ varies.

The Mathematical Appendix places these results in the context of recent work in
physical theory, i.e., Jarzynski's (1997) derivation of a nonequilibrium equality for
free energy differences.

Dynamic behavior of coevolutionary phenomena in our formulation can now be
derived via an Onsager approximation in the gradient of an iterated entropy-like
measure constructed from the iterated free energy Morse Function F via a Legendre
transform, in a curiously familiar manner (de Groot and Mazur 1984). That is, the
'entropy' gradient serves as a 'force' in system dynamics, driving the system from
low to higher 'entropy'.

We thus define such an 'entropy' in terms of the iterated free energy F as the
Legendre transform

$$S(Z) \equiv -F(Z) + ZdF(Z)/dZ \tag{2.6}$$

and take the time derivative of Z as defined—in first order—by the gradient in S:

$$\partial Z/\partial t \approx \mu \partial S/\partial Z = Zd^2F/dZ^2 \tag{2.7}$$

absorbing the 'diffusion coefficient' μ. Higher order—dimensionally consistent—
models are straightforward to implement, at the cost of considerable mathematical
burden.

It now becomes possible to introduce stochastic effects via the Ito Chain Rule
(Protter 2005) applied to the basic stochastic differential equation (SDE).

$$dZ_t = (Z_t d^2 F/dZ^2)dt + \sigma V(Z_t)dB_t \tag{2.8}$$

where the second term imposes what financial engineering calls volatility in the Brownian noise dB_t.

The central complication is that there is no single definition of stability for SDE's. Is the SDE stable in second order (i.e., variance)? In third order? Exponentially? Does it converge in probability? In the presence of Levy jumps? And so on. Each question has its own answer, often via the Ito Chain Rule. The circumstance is well recognized (e.g. Protter 2005; Appleby et al. 2008; Khasminskii 2012).

Imposing 'stability in order $Q(Z)$', application of the Ito Chain Rule leads to an averaged nonequilibrium steady state relation $< dQ_t >= 0$ that can be solved to find F as

$$F(Z) = -\int\int \frac{\sigma^2 V(Z)^2 d^2 Q/dZ^2}{2Z dQ/dZ} dZ\, dZ + C_1 Z + C_2 \tag{2.9}$$

Note that setting $< dQ_t >= 0$ generates a close analog to the asymptotic stochastic stationary states—the many 'fixed points'—of Dieckmann and Law (1996).

Again, these will be structured very exactly by the environmental imposition of 'stability' as a selection pressure.

Taking $V(Z) = Z$, a usual expression for volatility in financial engineering SDE's, leads to the relation for stochastic stability in second order—so that $Q(Z) = Z^2$—as

$$F(Z) = -\frac{\sigma^2}{4}Z^2 + C_1 Z + C_2 \tag{2.10}$$

This expression can now be used to characterize $g(Z)$ in Eq. (2.5), depending in detail on the partition function $\mathcal{H}(g(Z))$.

A direct implication of this result is that the 'simplicity' of enforced stability (in second order or otherwise) must be expressed through an enforced complexity in the biological regulatory apparatus necessary to implement it under selection pressure. This is a fundamental point.

This implication is, in fact, iterated if the underlying structure is, *prima face*, inherently unstable in the sense of the Data Rate Theorem of Control Theory (e.g. Nair et al. 2007). That theorem states that externally-supplied control information must be supplied at a rate greater than the inherently unstable system generates it's own 'topological information'. The canonical example is of a vehicle driven at high speed along a twisting roadway. The driver must impose steering, braking, and other control information at a rate greater than the road counterimposes it own 'twistiness' information, at the given vehicle speed. Biological examples include, for higher animals, stabilization of blood pressure, prevention of immune system self-attack, malignancy suppression, and restriction of the 'stream of consciousness' within 'riverbanks' useful to individual animals and their populations Wallace (2012a, 2022).

Then Eq. (2.8) becomes

$$dZ_t = \left(Z_t d^2 F / dZ^2 - M(Z_t)\right) dt + \sigma V(Z_t) dB_t \tag{2.11}$$

where $M(Z_t)$ is the rate at which regulatory control free energy must be imposed by external agents to stabilize the inherently unstable system.

For 'stability in order $Q(Z)$'—$< dQ_t >= 0$—the Ito Chain Rule gives

$$F(Z) =$$

$$\iint \frac{-\sigma^2 (V(Z))^2 \frac{d^2}{dZ^2} Q(Z) + 2M(Z) \frac{d}{dZ} Q(Z)}{2Z \frac{d}{dZ} Q(Z)} dZ\, dZ +$$

$$C_1 Z + C_2 \tag{2.12}$$

For stability in second order under 'ordinary' volatility $V(Z) = Z$, and most simply taking $M(Z) = M$ as a constant,

$$F(Z) = M \ln(Z) Z - MZ - \frac{\sigma^2 Z^2}{4} + C_1 Z + C_2 \tag{2.13}$$

which is significantly different from Eq. (2.10).

An indirect implication of this analysis is that stability under 'non-Brownian' colored noise—realistically, not having a uniform spectral density—will be significantly more complicated (Protter 2005; Hanggi and Jung 1995; Wallace 2016b). Indeed, noise color will likely prove to be an important part of both environmental effects and selection pressures that cannot be represented as information sources Y and $\mathscr{S}_P$ in Eq. (2.2). We do not address this question here, but Wallace (2016b) raises the caution

> ... [I]f evolutionary process can expect some phenomenon to evade selection pressure, it usually does, at some point and under some circumstances. Experimental and observational tests may, however, be subtle, if only for ideological reasons because we do not like to think that 'noise' itself might carry useful information. We may be missing something of no small biological significance.

Some of this is described in the Mathematical Appendix.

We provide two worked-out examples, at both ends of the scale spectrum, before continuing theoretical development.

2.4   Examples

First

Consider a system where the sum in the denominator of Eq. (2.4)—the partition function—can be approximated as an integral across a full set of source uncertainties, so that

$$\exp[-F/g] = \int_0^\infty \exp[-H/g]dH = g$$

$$g = \frac{-F}{W(n, -F)}$$

$$L \equiv \frac{\int_{H_0}^\infty \exp[-H/g]dH}{\int_0^\infty \exp[-H/g]dH} = \exp[-H_0/g] \qquad (2.14)$$

where F is the free energy analog, $g = g(Z)$ is the temperature analog from Eq. (2.4), and L represents a 'rate of reaction' for a triggering 'activation energy' threshold H_0 (Laidler 1987). $W(n, x)$ is, again, the Lambert W-function of order n, only real-valued for $n = 0, -1$ over limited ranges of x.

We next study the dynamics of the reaction rate L under uncertainty, specifically focusing on the nonequilibrium steady state corresponding to second-order stability, using Eq. (2.13)—here, taking $M(Z) = M$—and setting $M = 0, 0.15, H_0 = 1, C_1 = -1/2, C_2 = 7/2, \sigma = 0, 1$.

We use the W-function of order 0, which is real-valued only over the range $-F \geq -\exp[-1]$, a condition placing stability limitations on the system via complicated synergisms between Z, M, and σ at the particular boundary conditions defined by the integration constants C_1 and C_2.

The results, in Fig. 2.1, are consistent with—and indeed explicitly illustrate—the vagaries of the critical *RootOf* expression in Eq. (2.5).

In Fig. 2.1a the 'offset term' M in Eq. (2.13) is zero, and the reaction rate displays a kind of stochastic amplification roughly analogous to stochastic resonance. That is, $\sigma > 0$ triggers detection at a lower critical value of Z.

Figure 2.1b imposes an 'control offset' of $M = 0.15$. Remarkably, control free energy at a high enough rate M transforms the system to a signal transduction inverted-U Yerkes-Dodson pattern at $\sigma = 0$. Enough 'noise' σ, however, places the system in a more stable mode, *provided second-order stability can still be attained by regulation*. This is consistent with the observations of Appleby et al. (2008), who show that, under broad conditions, in one dimension, a 'noise' term can almost always be found that stabilizes an inherently unstable system, or, in two or more dimensions, destabilizes an inherently stable one.

At large enough values of σ, however, it is not possible to maintain second-order stability. This is most easily seen from Eq. (2.8), if we suppose an 'exponential' model for dZ/dt as $dZ/dt = Zd^2F/dZ^2 = \beta - \alpha Z$ and $V(Z) = Z$. Then the Ito Chain Rule applied to Z^2 finds the variance as

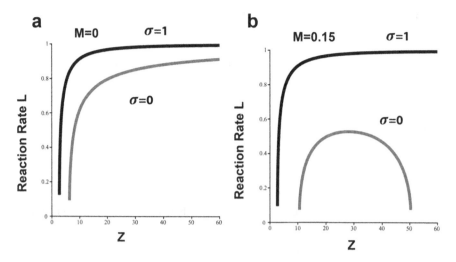

Fig. 2.1 (**a**) The 'offset term' M in Eq. (2.13) is zero, and the reaction rate displays a kind of stochastic amplification analogous to stochastic resonance. That is, $\sigma > 0$ triggers detection at a lower critical value of Z. (**b**) At a 'control offset' of $M = 0.15$ the system enters a classic signal transduction inverted-U mode for zero σ. Sufficient 'noise' then restores the inverted-J pattern. The limit on such dynamics is that, at sufficiently large σ, it becomes impossible to impose stability in second order

$$Var = \left(\frac{\beta}{\alpha - \sigma^2/2}\right)^2 - \left(\frac{\beta}{\alpha}\right)^2 \tag{2.15}$$

When $\sigma^2/2 \rightarrow \alpha$, second order stability fails, as it will at high enough σ for any system. Then Eq. (2.13) becomes impossible, and the analysis fails.

Second

The second model examines the other end of the complexity scale, a system split into only two levels, having source uncertainties $H_0 \pm \delta$, with δ again assumed small. The partition function/free energy relation is just

$$\exp[-F/g] = \exp[-(H_0 + \delta)/g] + \exp[-(H_0 - \delta)/g] \tag{2.16}$$

A clever expansion of F to second order in δ, and then solving the resulting expression for g, gives

$$g \approx -\frac{F - H - \sqrt{-2\ln(2)\,\delta^2 + F^2 - 2FH + H^2}}{2\ln(2)} \tag{2.17}$$

The reaction rate is calculated directly as

$$L(F) = \frac{\exp[-(H_0 + \delta)/g(F)]}{\exp[-(H_0 + \delta)/g(F)] + \exp[-(H_0 - \delta)/g(F)]} =$$

$$\frac{1}{1 + \exp[2\delta/g(F)]} \qquad (2.18)$$

Again assuming $M(Z) = M$ under ordinary volatility, so that

$$F = M \log(Z)Z - MZ - \sigma^2 Z^2/4 - Z/2 + 1$$

according to Eq. (2.13)—with $H_0 = 1$, $\delta = 0.1$, $\sigma = 0$, 1, $M = 0$, 1—calculation generates reaction rates in Fig. 2.2, analogous to—but significantly different from—Fig. 2.1.

Again, in Fig. 2.2a, with offset $M = 0$, $\sigma > 0$ amplifies reaction rate. In Fig. 2.2b the pattern again changes dramatically. Taking $M = 1$, reaction rate at $\sigma = 0$ first displays an inverted-U with increasing Z, but then a punctuated transition to a 'panic' state. At $\sigma = 1$, however, the reaction rate is again stabilized. To reiterate, as Appleby et al. (2008) indicate at some length, this is a familiar pattern as 'noise' terms can almost always be found that destabilize a stable system, or stabilize an unstable one.

Again, however, large enough σ makes it impossible to impose second-order stability without excessive consumption of essential resources, and the model fails.

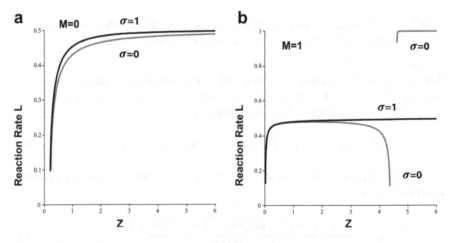

Fig. 2.2 Analog to Fig. 2.1 for reaction rate L, using the relations of Eq. (2.17) and (2.18). Here, $F = M \log(Z)Z - MZ - \sigma^2 Z^2/4 - Z/2 + 1$, with $H_0 = 1$, $\delta = 0.1$. (**a**) $M = 0$, $\sigma = 0$, 1. (**b**) $M = 1$, $\sigma = 0$, 1. Here, an inverted-U signal transduction at $\sigma = 0$ undergoes a phase transition to a panic mode, while 'noise' stabilizes the system. Again, however, sufficient noise makes second-order stability impossible, and the model fails

The Mathematical Appendix explores an intermediate-scale model that, as is typical, carries a significantly higher algebraic burden.

2.5   More Theory: Selection Pressure as Shadow Price

Central to gene frequency evolutionary models and their EES extensions is the idea of optimization on a 'fitness landscape' or on paths constrained within such a landscape (e.g. Gonzalez-Forero and Gardner 2021, 2022). More generally, one can seek to maximize some scalar index of critical function across a multi-component system constrained by resource rates and by the necessity of responding collectively to environmental signals or changes within time constraints. A central feature is an assumption that the rate at which an essential resource is delivered to component i is governed according to Eq. (2.7) as

$$\partial Z_i / \partial t = dS/dZ_i = f_i(Z_i) \tag{2.19}$$

The full system is seen as consisting of $i = 1, 2, \ldots, m$ cooperating units that must be individually supplied with resources at rates Z_i under the overall constraints of resources and time as $\sum_i Z_i = Z$, $\sum_i T_i = T$. We then optimize some critical scalar index of function across the full system under these constraints.

As an example, we take the sum of subcomponent reaction rates of Eqs. (2.14) and (2.18), assuming $dZ_i/dt = f(Z_i) \rightarrow 0$.

While more general approaches to optimization than simple Lagrangian are possible (e.g. Nocedal and Wright 2006), here we construct a general Lagrangian optimization across subcomponent reaction rates as

$$\mathcal{L} = \sum_i L_i + \lambda \left(Z - \sum_i Z_i \right) + \mu \left(T - \sum_i T_i \right)$$

$$\partial \mathcal{L} / \partial Z = \lambda$$

$$\partial L_i / \partial Z_i = \lambda$$

$$\partial L_i / \partial T_i = \mu \tag{2.20}$$

remembering in particular that $\partial Z_i / \partial T_i = f_i(Z_i)$.

For simple economic models, λ and μ represent the 'shadow prices' imposed on system dynamics by environmental constraints, in a large sense (Jin et al. 2008; Robinson 1993).

While many such formalisms are possible, elementary Lagrangian optimization outlines the basic mechanisms.

Algebraic manipulation—which will apply to virtually all possible scalar function measures—gives

$$f_i(Z_i) = \frac{\mu}{\lambda}$$

$$Z_i = f_i^{-1}(\frac{\mu}{\lambda}) \tag{2.21}$$

so that a monotonic function $f_i(Z_i)$ generates a monotonic dependence of Z_i on the shadow price ratio.

For example, setting $f_i(Z_i) = \beta_i - \alpha_i Z_i$—an 'exponential' model—produces the expression

$$Z_i = \frac{\beta_i - \mu/\lambda}{\alpha_i} \tag{2.22}$$

Sufficient environmental shadow price—selection pressure—will drive any component resource rate Z_i below what is needed for critical modular function.

Shadow prices characterize environmental burdens—selection pressures—leading to decline in the maximum possible level of resource delivery Z_{max} below some critical value Z_{crit}. Optimization becomes impossible, and we could then write both Z_{max} and Z_{crit} for a critical subsystem as appropriate functions of some integral of the environmental shadow price over the duration of acute selection pressure. Details will vary across organisms and environments, with optimization model and measures of λ, μ and Z, but the mechanism seems canonical.

Again, full consideration indicates that shadow prices will impose selection pressure in a similar manner for any possible scalar measure of fitness in this formulation.

Multivariate versions of the model, in which, for example, $Z \rightarrow Z^j$, $\lambda \rightarrow \lambda^j$, $j = 1, 2, \ldots$ and/or $T \rightarrow T^k$, $\mu \rightarrow \mu^k$, $k = 1, 2, \ldots$, lead in directions analogous to the classic 'fitness landscapes' described by Gavrilets (2010) and references therein. The shadow price parameters, and their relations, then characterize landscape topological roughness.

'Rough landscape' optimization bedevils a broad range of engineering, scientific, and practical enterprise, most particularly recent developments in machine learning and artificial intelligence, leading to far realms of topology, applied mathematics, and algorithm design.

2.6   Extending the Models

We have assumed systems as fully characterized by the single scalar parameter Z, mixing material resource/energy supply with internal and external flows of information. Following Wallace (2021a)—invoking techniques akin to Principal Component Analysis—it can be argued that there may often be more than one independent composite entity irreducibly driving system dynamics. Consequently, it may often be necessary to replace the scalar Z with an n-dimensional vector $\mathbf{Z}$

having orthogonal components that, together, account for most of the total variance in the rate of supply of essential resources. The basic dynamic equations then become considerably more complicated,

$$F(\mathbf{Z}) = -\log\left(\mathcal{H}(g(\mathbf{Z}))\right) g(\mathbf{Z})$$

$$S = -F + \mathbf{Z} \cdot \nabla_{\mathbf{Z}} F$$

$$\partial \mathbf{Z}/\partial t \approx \hat{\mu} \cdot \nabla_{\mathbf{Z}} S = f(\mathbf{Z})$$

$$-\nabla_{\mathbf{Z}} F + \nabla_{\mathbf{Z}}(\mathbf{Z} \cdot \nabla_{\mathbf{Z}} F) =$$

$$\hat{\mu}^{-1} \cdot f(\mathbf{Z}) \equiv f^*(\mathbf{Z})$$

$$\left(\left(\partial^2 F/\partial z_i \partial z_j\right)\right) \cdot \mathbf{Z} = f^*(\mathbf{Z})$$

$$\left(\left(\partial^2 F/\partial z_i \partial z_j\right)\right)|_{\mathbf{Z}_{nss}} \cdot \mathbf{Z}_{\mathbf{nss}} = \mathbf{0} \qquad (2.23)$$

F, g, $\mathcal{H}$, and S are scalar functions, while $\hat{\mu}$ is an n-dimensional square matrix of diffusion coefficient analogs.

The expression $((\partial F/\partial z_i \partial z_j))$ is an n-dimensional square matrix of second partial derivatives, and $f(\mathbf{Z})$ is a vector function. The last relation imposes a deterministic nonequilibrium steady state condition, i.e. $f^*(\mathbf{Z}_{nss}) = \mathbf{0}$. This may or may not be inherently stable.

Taking a particularly simple approach, $\mathcal{H}(g(\mathbf{Z})) \to g(\mathbf{Z})$, with $\mathbf{Z}(t) \to \mathbf{Z}_{nss}$.

This is an overdetermined system of partial differential equations (Spencer 1969) if $n \geq 2$. Indeed, for a general $f^*(\mathbf{Z})$, the system is inconsistent, resulting in as many as n different expressions for $F(\mathbf{Z})$, and hence the same number of temperature-analog measures.

This inference should not be surprising. The fifth expression in Eq. (2.23), where $f^*(\mathbf{Z}) \neq \mathbf{0}$, represents, most generally, a multi-component, cross-interacting, cross-talking, system that, if acting in real time, will almost always be far indeed from any steady state, equilibrium or nonequilibrium.

Such systems will almost always be inherently unstable, requiring constant input of control information that will necessarily lag perturbation. Such a structure should not, in fact, be characterizable by a single cognitive temperature-analog g. Each cognitive component, if the system is far from a steady state, should be expected to have its own g-value, in addition to structure imposed by the multidimensional nature of $\mathbf{Z}$.

The vector function $f(\mathbf{Z})$ becomes the basis for extension of Eq. (2.8) as a multidimensional SDE having the form

$$d\mathbf{Z}_t = f(\mathbf{Z}_t)dt + \rho(\mathbf{Z}_t)d\mathbf{B}_t \qquad (2.24)$$

where $d\mathbf{B}_t$ represents n-dimensional Brownian noise.

This development, where the dimension is $n \geq 2$, invokes the world of Appleby et al. (2008), in which it is almost always possible to find a functional form $\rho(\mathbf{Z})$

that destabilizes an inherently stable function $f(\mathbf{Z})$, or—by contrast—stabilizes an inherently unstable one.

In addition, as Wallace (2021a) shows, higher dimensional systems suffer the burdens of Lie Group methods in the study of their deterministic dynamics.

Likewise, as discussed above and shown in the Mathematical Appendix, Eqs. (2.9) and (2.24) can be extended to systems with colored noise, via the Dolean-Dade exponential, contingent on no small algebraic overhead (Protter 2005; Wallace 2016b).

The case $M(Z_t) = M Z_t$ in Eq. (2.11) is of particular interest for second order stability under ordinary volatility, i.e., $V(Z) = Z$:

$$F(Z) = \frac{Z^2}{2}\left(-\frac{\sigma^2}{2} + M\right) + C_1 Z + C_2 \tag{2.25}$$

generating obvious equivalence classes for M and σ defining a groupoid whose 'symmetry-breaking' can be associated with generalized phase transformations.

Further extensions of theory are possible, if not entirely straightforward. The Mathematical Appendix to Wallace (2021b) expands both the very idea of 'entropy' and the Onsager gradient model defining system dynamics, to higher orders. This can be done via formal power series algebraic ring structures, following the lead of Jackson et al. (2017), who see the Legendre transform as mapping the coefficients of one formal power series into the coefficients of another formal power series, modulo dimensional restrictions and consistency. One implication of this approach is that the 'spectra' of the formal power series coefficients characterize essential features of the systems under study, and will likely display particular dynamics under changing patterns of selection pressure and affordance.

2.7   Discussion

This work outlines a modestly comprehensive and 'mathematically elegant' expansion of the Extended Evolutionary Synthesis, expressed in terms of statistical models derived from the asymptotic limit theorems of information and control theories and abductions from related formalisms. While gene frequencies, and larger-scale direct-heritage mechanisms, do not dominate, they do not go away. Rather, as the 'message' in the transmission of genetic and other heritage information, they become 'words' emitted by a particular heritage information source. This source, however, must interact with others representing embedding ecosystems, niches, cognition modes—each with its associated regulatory agencies—characteristic large deviations, and structured selection pressures.

All of this occurs within the context of a powerful stochastic burden in which requirement for stability at a particular order—codified in Eqs. (2.9) and (2.12)—actually sets nonequilibrium steady states, as based on Onsager models in which time derivatives of essential parameters are determined by parameter gradients in an entropy analog built from a free energy construct.

Notably, the draconian simplicity of enforced stochastic stability—the 'stability in order Q' of Eqs. (2.9) and (2.12)—must inevitably and inversely be implemented by a necessary explosion of the complex regulatory machineries necessary to impose it, again, at and across various scales and levels of organization. These will range from the cellular to the institutional.

Note that, in this formulation, there can always be additional heritage information sources beyond genetic. These may be biochemical epigenetic or, in higher animals, social, institutional, and cultural (e.g. Jablonka and Lamb 1998; Jaenisch and Bird 2003; Avital and Jablonka 2000; Heine 2001).

Comprehensive formulations in terms of the asymptotic limit theorems of probability theory permit construction of both mathematical models and statistical tools whose most central utility is providing benchmarks against which to compare observational and experimental data (Pielou 1977, 1981). Such comparison encourages interplay between theory and data, contingent on a rigid determination not to fossilize scientific enterprise within the amber of any particular mathematical structure, however elegant.

Indeed, some do not find that gene frequency models embody much notable mathematical elegance: *de gustibus non disputandum est*.

The reader may not agree with the particular approach taken here, but this work does suggest that the EES can indeed be expanded and mathematically formalized in a comprehensive and 'elegant' manner. However, building useful regression model analogs from any such theory—and validating their realms of utility—remains typically challenging.

Most particularly, as (Pielou 1977, 1981) notes, the utility of mathematical models in the study of complex biological and ecological phenomena arises from a dynamic interplay between model-based suggestions and data-based conclusions, a relation in which models and statistical tools, however elegant, must always play a subordinate role. That being said, such models, and the statistical tools resulting from them, can serve as useful benchmarks against which to compare observational and experimental data.

2.8   Mathematical Appendix

An Intermediate Scale Model

Here we study, rather than the two levels of Fig. 2.2, the case of two sets of N levels each, distributed randomly above and below a mean level H_0.

The partition function F becomes

$$\exp[-F/g(Z)] = \sum_{j=1}^{N} \exp[-(H_0 + \delta_j)/g(Z)] + \sum_{i=1}^{N} \exp[-(H_0 - \delta_i)/g(Z)]$$

(2.26)

Taking $\exp[\delta/g(Z)] \approx 1 + \delta/g(Z) + \delta^2/2g(Z)^2$ for small δ and collecting terms—assuming the individual δ sum to zero about H_0—then

$$\exp[-F/g(Z)] \approx \exp[-H_0/g(Z)] \left(2N + \frac{\Delta}{2g(Z)^2}\right)$$

(2.27)

where $\Delta \equiv \sum_{k=1}^{2N} \delta_k^2$. That is, the sum is across all the individual δ^2.
Then

$$F = -\ln\left(\frac{4Ng(Z)^2 + \Delta}{2g(Z)^2}\right) g(Z) + H_0$$

(2.28)

and we set the right-hand side of Eq. (2.28) equal to the right-hand side of Eq. (2.13). This relation can be solved numerically for $g(Z)$. Figure 2.3 sets $N = \Delta = 100$, $M = 3$, $C_1 = -1$, $C_2 = 1$, $H_0 = 1$, $\sigma = 0, 1, 2$. As σ increases from zero, $g(Z)$ first rises monotonically with Z, but then displays the inverted-U signal transduction form. Calculation of the reaction rates, however, depends on the details of the distribution of states above and below H_0.

Colored Noise

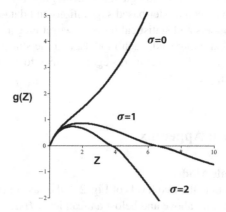

Fig. 2.3 We set the rhs of Eq. (2.28) equal to the rhs of Eq. (2.13) and solve numerically for $g(Z)$. Here, $N = \Delta = 100$, $C_1 = -1$, $C_2 = 1$, $M = 3$, $H_0 = 1$, $\sigma = 0, 1, 2$. As σ increases from zero, $g(Z)$ first rises monotonically with Z, but then displays an inverted-U Yerkes-Dodson signal transduction. Calculation of the reaction rate depends in detail on the distribution of states above H_0

Following Protter (2005), the quadratic variation $[X, Y]_t$ between two continuous stochastic processes can be written in terms of Ito stochastic integrals as

$$X_t Y_t = X_0 Y_0 + \int_0^t X_{s-} dY_s + \int_0^t Y_{s-} dX_s + [X, Y]_t \tag{2.29}$$

where $s-$ indicates left-continuity. For a discrete process the integrals are rewritten as sums to produce a 'simpler' expression.

For Brownian white noise dB_t, $[B_t, B_t] \propto t$. Colored noise will produce a different time dependence.

If an SDE can be written as $dX_t = X_{t-} dY_t$, where Y_t is a stochastic process, the solution is given by the Doleans-Dade exponential as

$$X_t \propto \exp\left(Y_t - \frac{1}{2}[Y, Y]_t\right) \tag{2.30}$$

By the mean value theorem, if

$$\frac{1}{2} d[Y, Y]/dt > dY/dt \tag{2.31}$$

then $X_t \to 0$, remembering that, for Brownian processes. $[Y, Y] \propto t$.

The argument can be directly, if not easily, extended to Levy jump-processes (Protter 2005).

In Parallel with Jarzynski

Somewhat surprisingly, it is possible to contrast and compare the cognitive phase transitions studied here with Jarzynski's (1997) classic derivation of a nonequilibrium equality for free energy differences.

The Second Law of thermodynamics famously states that, for *finite time* transitions between equilibrium states along any path, the average work needed will be greater than or equal to the change in free energy ΔF between the states:

$$< \mathcal{W} >\geq \Delta F \tag{2.32}$$

where the '$< >$' bracket indicates an average over an ensemble of measurements of the work $\mathcal{W}$. Jarzynski's elegant extension is the *exact* statement

$$< \exp[-\mathcal{W}/kT] >= \exp[-\Delta F/kT]$$
$$\Delta F = -kT \log\left(< \exp[-\mathcal{W}/kT] >\right) \equiv -kT \log\left(h(kT)\right) \tag{2.33}$$

where T is the absolute temperature and k an appropriate constant. Equation (2.32) emerges from the first of these expressions via the Jensen inequality $< \phi(x) >\geq \phi(< x >)$ for any convex function ϕ.

Inverting the perspective—thus making free energy more fundamental than temperature—*we can solve the Jarzynski relation for kT*, giving, as above, a 'Fisher zero' set of phase transition solutions as

$$kT = -\frac{\Delta F}{RootOf\left(e^X - h\left(-\frac{\Delta F}{X}\right)\right)} \tag{2.34}$$

where, again, X is a dummy variate in the RootOf construct generalizing the Lambert W-function.

That function, in fact, emerges directly if $h(kT)$ were imagined to have a strongly dominant term $a(kT)^b$. Then

$$kT \approx (-\Delta F)/[b\, W(n, -\Delta F a^{1/b}/b)] \tag{2.35}$$

Taking $n = 0, -1$, this expression will again be subject to 'Fisher zero' imaginary component phase transitions if ΔF exceeds appropriate ranges.

These results emphasize the considerable degree to which the dynamics of cognitive processes are likely to differ from the more familiar dynamics of physical—or even biophysical—phenomena. That is, cognitive systems are channeled by the asymptotic limit theorems of information and control theories (Dretske 1994), in addition to usual physical constraints, and this is most often a different world indeed.

More specifically, while, for physical systems, free energy dynamics may be dominated by 'ordinary' temperature changes across phase transitions, cognitive phase transitions—involving the 'cognitive temperature' $g(Z)$ of Eqs. (2.4) and (2.5)—normally occur at fixed physical temperature and are *driven by changes in rates of available free energies that include rates of information transmission*.

Singular characteristics of cognitive dynamics, in particular 'Yerkes-Dodson laws' and their generalizations, emerge from this fundamental difference.

A further divergence from physical theory emerges in our adaptation of Onsager's formalism for nonequilibrium thermodynamics, defining a free energy analog using the 'partition function' of the denominator of Eq. (2.4). This permits construction of an entropy-analog via a Legendre transform, and versions of the basic Onsager approach, i.e., time derivatives of driving parameters being taken as proportional to the divergence of entropy by those parameters. Multi-dimensional forms of these models, however, do not sustain 'Onsager reciprocal relations' since information sources are not microreversible: in English 'eht' does not have the same probability as 'the'. Indeed, for information sources in general, palindromes have vanishingly small probabilities.

That is, the topology of information sources is dominated by directed homotopy, leading to equivalence class groupoid symmetries and phase transitions in terms of broken groupoid symmetries. Indeed Cimmelli et al. (2014) describe similar matters afflicting nonlinear nanoscale phenomena that appear consonant with the fluctuation-dissipation studies cited above.

References

Adami, C., and N. Cerf, 2000. Physical complexity of symbolic sequences. Physica D 137:62–69.

Adami, C., C. Ofria, and T. Collier, 2000. Evolution of biological complexity. PNAS 97:4463–4468.

Appleby, J., X. Mao, and A. Rodkina, 2008. Stabilization and destabilization of nonlinear differential equations by noise. IEEE Transactions on Automatic Control 53:68–69.

Atlan H., and I. Cohen, 1998. Immune information, self-organization, and meaning. International Immunology 10:711–717.

Avital, E., and E. Jablonka, 2000. *Animal Traditions: Behavioral Inheritance in Evolution.* Cambridge: Cambridge University Press.

Bazil J., G. Buzzard, and A. Rundell, 2020. Modeling mitochondrial bioenergetics with integrated volume dynamics. PLoS Computational Biology 6:e1000632.

Brown, R., 1992. Out of line. Royal Institute Proceedings 64:207–243.

Cayron, C., 2006. Groupoid of orientational variants. Acta Crystalographica Section A A62:21040.

Champagnat, N., R. Ferriere, and S. Meleard, 2006. Unifying evolutionary dynamics: From individual stochastic process to macroscopic models. Theoretical Population Biology 69:297–321.

Chvaja, R., 2020. Why did memetics fail? Comparative case study. Perspectives in Science 28:542–570.

Cimmelli, V., A. Sellitto, and D. Jou, 2014. A nonlinear thermodynamic model for a breakdown of the Onsager symmetry and the efficiency of thermoelectric conversion in nanowires. Proceedings of the Royal Society A 470:20140265.

Cohen, I., and D. Harel, 2007. Explaining a complex living system: dynamics, multi-scaling and emergence. Journal of the Royal Society Interface 4:175–182.

Cover, T., and J. Thomas, 2006. *Elements of Information Theory*, 2nd ed. New York: Wiley.

de Groot, S., and P. Mazur, 1984. *Nonequilibrium Thermodynamics.* New York: Dover.

Dembo, A., and O. Zeitouni, 1998. *Large Deviations and Applications*, 2nd ed. New York: Springer.

Dieckmann, U., and R. Law, 1996. The dynamical theory of coevolution: a derivation from stochastic ecological processes. Journal of Mathematical Biology 34:579–612.

Dolan, B., W. Janke, D. Johnston, and M. Stathakopoulos, 2001. Thin Fisher zeros. Journal of Physics A 34:6211–6223.

Dretske, F., 1994. The explanatory role of information. Philosophical Transactions of the Royal Society A 349:59–70.

EES, 2020. https://extendedevolutionarysynthesis.com/resources/recommended-reading/.

Feynman, R. 2000. *Lectures on Computation.* New York: Westview Press.

Fisher, M., 1965. *Lectures in Theoretical Physics*, Vol. 7. Boulder: University of Colorado Press.

Gavrilets, S., 2010. High-dimensional fitness landscapes and speciation. In *Evolution: The Extended Synthesis*, ed. Massimo Pigliucci and Gerd B. Müller. Cambridge: MIT Press.

Glazebrook, J., and R. Wallace, 2012. 'The Frozen Accident' as an evolutionary adaptation: a rate distortion theory perspective on the dynamics and symmetries of genetic coding mechanisms. Informatica 36:53–73.

Gonzalez-Forero, M., and A. Gardner, 2021. A mathematical framework for evo-devo dynamics. Preprint, bioRxiv. https://doi.org/10.1101/2021.05.17.444499.

Gonzalez-Forero, M., and A. Gardner, 2022. How development affects evolution. Preprint, bioRxiv https://doi.org/10.1101/2021.10.20.464947.

Hanggi, P., and P. Jung, 1995, Colored noise in dynamical systems. Advances in Chemical Physics 89:239–326.

Heine, S., 2001. Self as cultural product: an examination of East Asian and North American selves. Journal of Personality 69:881–906.

Jablonka, E., and M. Lamb, 1998. Epigenetic inheritance in evolution. Journal of Evolutionary Biology 11:159–183.

Jackson, D., A. Kempf, and A. Morales, 2017. A robust generalization of the Legendre transform for QFT. Journal of Physics A 50:225201.

Jaenisch, R., and A. Bird, 2003. Epigenetic regulation of gene expression: how the genome integrates intrinsic and environmental signals. Nature Genetics Supplement 33:245–254.

Jarzynski, C., 1997. Nonequilibrium equality for free energy differences. Physical Review Letters 78:2690–2693.

Jin, H., Z. Hu, and X. Zhou, 2008. A convex stochastic optimization problem arising from portfolio selection. Mathematical Finance 18:171–183.

Khasminskii, R., 2012. *Stochastic Stability of Differential Equations*. New York: Springer.

Khinchin, A., 1957. *Mathematical Foundations of Information Theory*. New York: Dover.

Laidler, K., 1987. *Chemical Kinetics*, 3rd ed. New York: Harper and Row.

Laland, K., F. Odling-Smee, and M. Feldman, 1999. Evolutionary consequences of niche construction and their implications for ecology. PNAS 96:10242–10247.

Laland, K., T. Uller, M. Feldman, K. Sterelny, G. Muller, A. Moczek, E. Jablonka, and J. Odling-Smee, 2014. Does evolutionary theory need a rethink? Nature 514:163–164.

Laland, K., T. Uller, M. Feldman, K. Sterelny, G. Muller, A. Moczek, E. Jablonka, and J. Odling-Smee, 2015. The extended evolutionary synthesis: its structure, assumptions and predictions. Proceedings of the Royal Society B 282:20151019.

Landau, L., and E. Lifshitz, 1980. *Statistical Physics*, 3rd ed. New York: Elsevier.

Love, A., 2010. Rethinking the structure of evolutionary theory for an extended synthesis. In *Evolution: The Extended Synthesis*, ed. Massimo Pigliucci and Gerd B. Müller. Cambridge: MIT Press.

Maturana, H., and F. Varela, 1980. *Autopoiesis and Cognition: The Realization of the Living*. Boston: Reidel.

Muller, G., 2017. Why an extended evolutionary synthesis is necessary. Royal Society Interface 7:20170015.

Nair, G., F. Fagnani, S. Zampieri, and R. Evans, 2007. Feedback control under data rate constraints: an overview. Proceedings of the IEEE, 95:108–137.

Nocedal, J., and S. Wright, 2006. *Numerical Optimization*, 2nd ed. New York: Springer.

Odling-Smee, F., K. Laland, and M. Feldman, 2003. Niche construction: the neglected process in evolution. *Monographs in Population Biology*, Vol. 37. Princeton: Princeton University Press.

Ofria, C., C. Adami, and T. Collier, 2003. Selective pressures on genomes in molecular evolution. Journal of Theoretical Biology 222:477–483.

Pettini, M., 2007. *Geometry and Topology in Hamiltonian Dynamics and Statistical Mechanics*. New York: Springer.

Pielou, E., 1977. *Mathematical Ecology*. New York: John Wiley and Sons.

Pielou, E., 1981. The usefulness of ecological models: a stock-taking. Quarterly Review of Biology 56:17–31.

Pigliucci, M., and G. Muller, 2010. *Evolution: The Extended Synthesis*. Cambridge: MIT Press.

Protter, P., 2005. *Stochastic Integration and Differential Equations: A New Approach*, 2nd ed. New York: Springer.

Robinson, S., 1993. Shadow prices for measures of effectiveness II: general model. Operations Research 41:536–548.

Ruelle, D., 1964. Cluster property of the correlation functions of classical gases. Reviews of Modern Physics April:580–584.

Spencer, D., 1969. Overdetermined systems of linear partial differential equations. Bulletin of the American Mathematical Society 75:179–239.

Wallace, R., 2002. Adaptation, punctuation, and rate distortion: non-cognitive 'learning plateaus' in evolutionary process. Acta Biotheoretica 50:101–116.

Wallace, R., 2009. Metabolic constraints on the eukaryotic transition. Origins of Life and Evolution of Biospheres 39:165–176.

Wallace, R., 2010. Expanding the modern synthesis. Comptes Rendus Biologies 333:701–709.

Wallace, R., 2011a. On the evolution of homochirality. Comptes Rendus Biologies 334:263–268.

Wallace, R., 2011b. A new formal approach to evolutionary processes in socioeconomic systems. Journal of Evolutionary Economics 23:1–15.

Wallace, R., 2011c. A formal approach to evolution as self-referential language. BioSystems, 106:36–44.

Wallace, R., 2012a. Consciousness, crosstalk, and the mereological fallacy: an evolutionary perspective. Physics of Life Reviews 9:426–453.

Wallace, R., 2012b. Metabolic constraints on the evolution of genetic codes: did multiple 'preaerobic' ecosystem transitions entrain richer dialects via Serial Endosymbiosis? Transactions on Computational Systems Biology XIV, LNBI 7625:204–232.

Wallace, R., 2013a. A new formal perspective on 'Cambrian explosions'. Comptes Rendus Biologies 337:1–5.

Wallace, R., 2013b. A new formal approach to evolutionary process in socioeconomic systems. Journal of Evolutionary Economics 23:1–15.

Wallace, R., 2013c, Cognition and biology: perspectives from information theory, Cognitive Processing, 15:1–12.

Wallace, R., 2016a. The metabolic economics of environmental adaptation. Ecological Modelling 322:48–53.

Wallace, R., 2016b. Subtle noise structures as control signals in high-order biocognition. Physics Letters A 380:726–729.

Wallace, R., 2021a. How AI founders on adversarial landscapes of fog and friction. Journal of Defense Modeling and Simulation. https://doi.org/10.1177/1548512920962227.

Wallace, R., 2021b. Toward a formal theory of embodied cognition. BioSystems 202:104356.

Wallace, R., 2022. *Consciousness, Cognition and Crosstalk: The evolutionary exaptation of nonergodic groupoid symmetry-breaking*. New York: Springer.

Wallace, R., and R.G. Wallace, 1998. Information theory, scaling laws and the thermodynamics of evolution. Journal of Theoretical Biology 192:545–555.

Wallace, R., and R.G. Wallace, 1999. Organizations, organisms and interactions: an information theory approach to biocultural evolution. BioSystems, 51:101.

Wallace, R., and D. Wallace, 2008. Punctuated equilibrium in statistical models of generalized coevolutionary resilience: how sudden ecosystem transitions can entrain both phenotype expression and Darwinian selection. Transactions on Computational Systems Biology, IX, LNBI 5121, 23–85.

Wallace, R., and D. Wallace, 2009. Code, context and epigenetic catalysis in gene expression. Transactions on Computational Systems Biology, XI, LNBI 5750:283–334.

Wallace, R.G., and R. Wallace, 2009. Evolutionary radiation and the spectrum of consciousness. Consciousness and Cognition 18:160–167.

Wallace, R., and D. Wallace, 2011. Cultural epigenetics and the heritability of complex diseases. Transactions on Computational Systems Biology XIII, LNBI 6575: 131–170.

Wallace, R., and D. Wallace, 2016. *Gene Expression and Its Discontents: The Social Production of Chronic Disease*, 2nd ed. New York: Springer.

Wallace, R., D. Wallace, and R.G. Wallace, 2009. *Farming Human Pathogens: Ecological Resilience and Evolutionary Process*. New York: Springer.

Weinstein, A., 1996. Groupoids: unifying internal and external symmetry. Notices of the American Mathematical Association 43:744–752.

Wu, F., F. Yang, K. Vinnakota, and D. Beard, 2007. Computer modeling of mitochondrial tricarboxylic acid cycle, oxidative phosphorylation, metabolite transport, and electrophysiology. Journal of Biological Chemistry 282:24525–24537.

Chapter 3
On Regulation

3.1 Introduction

Maturana and Varela (1980) hold that all organisms, even those without a nervous system, are cognitive, and that living per se is a process of cognition. This seems indeed compelling, but is, at best, only half the story. Many—indeed, virtually all—real-world, real-time cognitive processes, much like driving on a twisting, pot-holed roadway, are inherently unstable, and must be heavily regulated by a 'driver' providing control information at a rate greater than the underlying road generates it's own 'topological information' (e.g., Nair et al. 2007). The human 'stream of consciousness' is restrained by social and cultural as well as more common neuropsychological 'riverbanks' for successful confrontation and circumvention of powerful selection pressures at both individual and group scales of organization (e.g., Wallace 2022a). Blood pressure rises and falls according to need, but must be held within critical limits for individual survival. Similarly, immune systems conduct routine cellular maintenance along with cancer and pathogen control, but must be restrained from attacking self-tissues (Atlan and Cohen 1998). And so on.

Most directly, cognition implies choice, choice reduces uncertainty, and any reduction of uncertainty necessarily implies the existence of an information source 'dual' to the cognitive process under study (Atlan and Cohen 1998; Wallace 2005, 2012, 2020, 2022a,b,c). We take such information sources to be piecewise adiabatically stationary. This means that on the 'piece' of a developmental trajectory, the system is close enough to non ergodic stationary for the appropriate asymptotic limit theorems of information theory to work sufficiently well. One possible analog is the Born-Oppenheimer approximation in molecular quantum mechanics in which nuclear motions are assumed to be slow enough for electron structures to equilibrate around them.

The asymptotic limit theorems of information theory are related to control theory by the Data Rate Theorem (e.g., Nair et al. 2007). Consider the first-order linear control system model of Fig. 3.1.

R. Wallace, *Essays on the Extended Evolutionary Synthesis*, SpringerBriefs in Evolutionary Biology, https://doi.org/10.1007/978-3-031-29879-0_3

Fig. 3.1 A simplified model
of an inherently unstable
system stabilized by a control
signal U_t. X_t is the system
state at time t and W_t the
'noise' perturbing the system
at time t

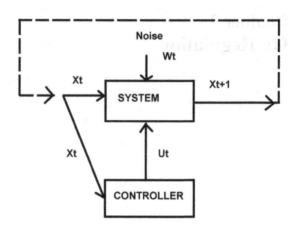

Let x_t be a first n-dimensional vector of system parameters at time t, assumed
near a nonequilibrium steady state. The state at time $t + 1$ is then approximated by
a first-order relation as

$$x_{t+1} \approx \mathbf{A}x_t + \mathbf{B}u_t + W_t \tag{3.1}$$

Here, $\mathbf{A}$ and $\mathbf{B}$ are fixed n-dimensional square matrices, u_t is a vector of control
information, and W_t is an n-dimensional vector of Brownian white noise.

The Data Rate Theorem (DRT) holds that if $\mathcal{H}$ is a rate of control information
sufficient to stabilize an inherently unstable control system, then it must be greater
than a minimum value $\mathcal{H}_0$ determined as

$$\mathcal{H} > \mathcal{H}_0 \equiv \log[\| \det[\mathbf{A}^m] \|] \tag{3.2}$$

where det is the determinant of the subcomponent $\mathbf{A}^m$—with $m \leq n$—of the matrix
$\mathbf{A}$ having eigenvalues ≥ 1. $\mathcal{H}_0$ defines the rate at which the unstable system itself
generates 'topological information'—the waviness and pot-hole frequencies of the
road at the chosen vehicle speed.

If this inequality is violated—if the control information rate falls below the system's topological information rate—stability fails. Then blood pressure skyrockets,
the immune system attacks self-tissues, conscious animals suffer hallucinations, and
the vehicle crashes.

Here, we take a somewhat unusual perspective that permits both generalizations
and applications, deriving control theory's Data Rate Theorem from information
theory's Rate Distortion Theorem. The approach finds surprisingly wide application
across stochastic influences on gestalt cognition rates under differing patterns of
regulatory constraint.

3.2 Theory

A Correspondence Reduction

We begin by briefly reprising Einstein's 1905/1956 treatment of Brownian motion for $N \gg 1$ particles, founded on the simplest possible diffusion relation

$$\partial\rho(x,t)/\partial t = \mu\partial^2\rho(x,t)/\partial x^2 \tag{3.3}$$

where $\rho(x)$ is the particle density at position x and μ the diffusion coefficient. This has an elementary solution via the Normal Distribution,

$$\rho(x,t) = \frac{N}{\sqrt{4\pi\mu t}} \exp[-x^2/(4\mu t)] \tag{3.4}$$

leading to an expression for the time dependence of averaged position

$$\sqrt{<x^2>} \propto \sqrt{t} \tag{3.5}$$

Analogous treatment of information system dynamics—essentially, a correspondence reduction from Shannon to Einstein—is surprisingly straightforward, modulo a recognition of information as a form of free energy, and not, in itself, an 'entropy', in spite of its formal 'entropy-like' mathematical representation in stationary, ergodic systems (Khinchin 1957; Feynman 2000).

More specifically, Bennett, as represented by Feynman (2000), constructs an ideal machine that extracts free energy—energy that can do work—from an information source. From another direction, information theory's Rate Distortion Theorem (Cover and Thomas 2006) determines the minimum channel capacity, written as $R(D)$, that is necessary to ensure that a signal transmitted along a noisy channel is received with an average distortion less than or equal to a particular scalar measure D. The typical worst case is the so-called Gaussian channel for which, taking the square distortion measure (Cover and Thomas 2006),

$$R(D) = \frac{1}{2}\log_2(\sigma^2/D), \quad D \leq \sigma^2$$

$$R(D) = 0, \quad D > \sigma^2 \tag{3.6}$$

Following Feynman (2000), here we explicitly identify information as another form of free energy. It is then possible to construct an entropy S in a standard manner using the Legendre transform, so that

$$S \equiv -R(D) + DdR/dD \tag{3.7}$$

The next step is to impose a first-order nonequilibrium thermodynamics Onsager approximation (de Groot and Mazur 1984), so that the rate of change of the scalar parameter D, the average distortion, is given as the gradient of S in D:

$$dD/dt \propto dS/dD = Dd^2R/dD^2 \tag{3.8}$$

where we have absorbed the diffusion coefficient into the formalism.

Solving this differential equation, using Eq. (3.6),

$$D(t) \propto \sqrt{t} \tag{3.9}$$

the same result as for simple Brownian motion, suggesting a canonical direction for the extension of theory.

Suppose a 'control free energy' is actively imposed on the system of interest at some rate $M(\mathscr{H})$, depending on a control information rate $\mathscr{H}$. The condition for stability is

$$dD/dt = \mu Dd^2R/dD^2 - M(\mathscr{H}) \leq 0$$

$$M(\mathscr{H}) \geq \mu Dd^2R/dD^2 \geq 0$$

$$\mathscr{H} \geq \mathscr{H}_0 \equiv \max\{M^{-1}(\mu Dd^2R/dD^2)\}\} \tag{3.10}$$

Here, max is taken as the maximum value, noting that, if M is monotonic increasing, so is the inverse function M^{-1}.

By convexity of the Rate Distortion Function (Cover and Thomas 2006), $d^2R/dD^2 \geq 0$.

In the absence of further noise, at the Gaussian channel's nonequilibrium steady state, $dD/dt \equiv 0$, and $D \propto 1/M(\mathscr{H})$.

A typical scenario might involve continuous radar or lidar illumination of a moving target with transmitted energy rate M, defining a maximum D inversely proportional to M. If illumination fails, then D will increase as $\sqrt{t}$, representing classic diffusion from the original tracking trajectory.

We have, in the last expression of Eq. (3.10), replicated something in the direction of Eq. (3.2), but there is far more structure to be uncovered.

Expanding the Model

The Rate Distortion Function $R(D)$ is always convex in D, so that $d^2R/dD^2 \geq 0$ (Cover and Thomas 2006; Effros et al. 1994). Following Effros et al. (1994) and Shields et al. (1978), for a nonergodic process, where the cross-sectional mean is not the time-series mean, the Rate Distortion Function can be calculated as an average across the RDF's of the ergodic components of that process, and can thus be expected to remain convex in D.

It is possible to reconsider and reinterpret Fig. 3.1, examining the 'expected transmission' of a signal $X_t \rightarrow \hat{X}_{t+1}$—now in the presence of an added 'noise' Ω—but received as a de-facto signal X_{t+1}. That is, there will be some deviation between what is ordered and what is observed, measured under a scalar distortion metric as $d(\hat{X}_{t+1}, X_{t+1})$, now averaged as

$$D = \sum Pr(\hat{X}_{t+1})d(\hat{X}_{t+1}, X_{t+1}) \tag{3.11}$$

We take Pr as the probability of $\hat{X}_{t+1}$ and recognize that the 'sum' can represent an appropriate generalized integral.

We have built an adiabatically, piecewise stationary information channel for the control system and are now able to invoke a Rate Distortion Function $R(D)$.

Following Eq. (3.10), it is now possible to write a general stochastic differential equation (Protter 2005)

$$dD_t = \left(D_t [d^2 R/dD^2]_t - M(\mathscr{H}) \right) dt + \Omega h(D_t) dB_t \tag{3.12}$$

$\Omega h(D)$ is taken as a 'volatility' under Brownian noise dB_t.

An average $< D >$ can be calculated from the relation $< dD_t >= Dd^2 R/dD^2 - M(\mathscr{H}) = 0$, so that something like Eq. (3.10) holds. However, stability of systems characterized by stochastic differential equations is a far more complex than for deterministic systems. More specifically, we are concerned with stability under stochastic volatility, and there are numerous definitions possible.

Here, we calculate the nonequilibrium steady state properties of a general function $Q(D_t)$ given Eq. (3.12)—$< dQ_t >= 0$—using the Ito Chain Rule (Protter 2006). Direct calculation produces

$$\left(\mu Dd^2 R/dD^2 - M(\mathscr{H}) \right) dQ/dD +$$

$$\frac{\Omega^2}{2} h(D)^2 d^2 Q/dD^2 = 0 \tag{3.13}$$

Explicitly solving for $M(\mathscr{H})$ gives

$$M(\mathscr{H}) = \frac{\Omega^2 (h(D))^2 \frac{d^2}{dD^2} Q(D)}{2 \frac{d}{dD} Q(D)} + D \frac{d^2}{dD^2} R(D) \tag{3.14}$$

This relation can, in turn, be solved in terms of M^{-1} for $\mathscr{H}$, since, by convexity, $d^2 R/dD^2 \geq 0$.

The results extend the Data Rate Theorem beyond the simple 'control information' rate.

Assuming $h(D) = D$, which is the classic volatility of financial engineering, and setting $Q(D) = D^2$ gives the condition for second order stability for a Gaussian channel—$R(D) = (1/2) \log_2(\sigma^2/D)$—as

$$M(\mathscr{H}) = \frac{\Omega^2}{2} D + \frac{1}{D \log(4)} \geq \frac{\Omega}{\sqrt{\log(2)}} \tag{3.15}$$

The last inequality represents the minimization $dM/dD = 0$.

Another approach takes the control signal $\mathscr{H}$ itself as the fundamental distortion measure between intended and observed behaviors, so that Eq. (3.11) becomes

$$H \equiv \sum Pr(\hat{X}_{t+1}) \mathscr{H}(\hat{X}_{t+1}, X_{t+1}) \tag{3.16}$$

where $\mathscr{H}$ is the rate of control information needed to stabilize the inherently unstable system and the other factors are as above. It is now possible to carry through the analysis as driven by Eq. (3.12), but with D replaced by H and the rate of control free energy is taken as M, again producing Eqs. (3.14) and (3.15).

Taking the Legendre transform and Onsager approximations of Eqs. (3.7) and (3.8), and through the SDE of Eq. (3.12), in Eq. (3.14) we derived something much like Eq. (3.2). One important but counterintuitive constraint is that, since information transmission is not microreversible—for example, in English the term 'the' has much higher probability than 'eht' and in Chess, it is impossible to uncheckmate a King—there can be no 'Onsager reciprocal relations' in multidimensional variants of Eq. (3.8).

Moving on: An Iterated Free Energy

Here, we reconsider and extend some results of Wallace (2022b).

Consideration of ensembles of possible processes markedly expands the perspective, to understate the matter:

1. Control dynamics—the set of 'Data Rate Theorems' available to a system— is context-dependent. Different 'road conditions' require different analogs to, or forms of, the DRT. More specifically, the transmission of control messages as represented in Eq. (3.11) involves equivalence classes of possible control sequences $X \equiv \{X_t, X_{t+1}, X_{t+2}, ...\}$. For example, driving a particular stretch of road slowly on a dry, sunny morning is a different 'game' than driving it at high speed during a midnight snowstorm.
2. It is necessary to characterize the set of all possible such 'games' in terms of equivalence classes G_j of the control path sequences X. Each class is then associated with a particular cognitive 'game' represented by a 'dual' information source having uncertainty

$$H_{G_j} \equiv H_j$$

that will vary across the complexity of the 'games' being played. Think of the sunny morning vs snowstorm driving example.

The idea of a 'dual' information source arises from the necessity of making behavioral choices in game-playing.

Choice reduces uncertainty, and the reduction of uncertainty implies existence of an information source dual to the cognitive process. The inference is direct and unambiguous. Different 'games' are associated with different 'languages' of play involving such choice.

3. In this model, $R(D)$, associated with a particular 'vehicle', now plays a different role. Here, we impose R via temperature analog $g(R)$ in an iterated development, through the pseudoprobability

$$P_j = \frac{\exp[-H_i/g(R)]}{\sum_j \exp[-H_i/g(R)]} \tag{3.17}$$

where, unlike 'ordinary' statistical thermodynamics, the 'temperature' $g(R)$ will have to be calculated from first principles.

4. The denominator in Eq. (3.17) represents an analog to the standard statistical mechanical partition function, and can be used to derive an iterated 'free energy' analog F as

$$\exp[-F/g(R)] = \sum_k \exp[-H_k/g(R)] \equiv A(g(R))$$

$$F(R) = -\log[A(g(R))]g(R) \tag{3.18}$$

5. It then becomes possible to define a new 'entropy' via a Legendre transform on F, and to again impose a dynamic relation like that of the first-order Onsager treatment of nonequilibrium thermodynamics (de Groot and Mazur 1984):

$$S(R) \equiv -F(R) + RdF/dR$$

$$\partial R/\partial t \approx dS/dR = Rd^2F/dR^2 = f(R)$$

$$F(R) = \int \frac{f(R)}{R} \, d[R, R] + C_1 R + C_2$$

$$g(R) = \frac{-F(R)}{RootOf \, (\exp[Z] - A(-F/Z))} \tag{3.19}$$

This last step is not as simple as it looks. There are a number of important implications:

- $f(R)$ represents the 'friction' inherent to any control system, the time it takes to react, for example, $dR/dt = f(R) = \beta - \alpha R$, $R(t) \to \beta/\alpha$.
- The RootOf construction generalizes the Lambert W-function, which can be seen by carrying through a calculation setting $A(g(R)) = g(R)$.

- Since the 'RootOf' construction may have complex number solutions, the temperature analog $g(R)$ now imposes 'Fisher Zeros' analogous to those associated with 'ordinary' phase transition in physical systems (Dolan et al. 2001; Fisher 1965; Ruelle 1964, Sec. 5). Phase transitions in cognitive process range from the punctuated onset of conscious signal detection to Yerkes-Dodson 'arousal' dynamics (Wallace 2022a).
- The set of equivalence classes G_j defines a groupoid (Weinstein 1996), and the Fisher Zeros represent groupoid symmetry-breaking for cognitive phenomena that is analogous to, but markedly different from, the group symmetry-breaking associated with phase transition in physical processes.
- As a result, groupoid symmetry-breaking phase transitions represent a significant extension of the basic control theory DRT.
- Another possible extension is via a 'reaction rate' model abducted from chemical physics (Laidler 1987). The 'reaction rate' L, taking the minimum possible source uncertainty across the 'game' played by the control system as H_0, is then

$$L = \frac{\sum_{H_j \geq H_0} \exp[-H_j/g(R)]}{\sum_k \exp[-H_k/g(R)]} \equiv \mathscr{L}(H_0, g(R)) \tag{3.20}$$

where the sums may be generalized integrals and the denominator can be recognized as $A(g(R))$ from Eq. (3.18).

Wallace (2022a) uses analogous models to derive a version of the Yerkes-Dodson 'inverted-U' model for cognition rate vs. arousal.

Fisher Zero analogs in $g(R)$ will also characterize cognition rate phase transitions.

- Extension of the theory to higher dimensions—writing $R(D, \mathbf{Z})$, where $\mathbf{Z}$ is an irreducible vector of resource rates—is subtle (Wallace 2022c, Ch. 5), as is extension to higher order versions of 'entropy' and of the Onsager entropy gradient models (Wallace 2021).
- The approach can be extended to stationary nonergodic systems—for which time averages are not ensemble averages—assuming that sequences can be broken into small high probability sets consonant with underlying forms of grammar and syntax, and a much larger set of low probability sequences not so consonant (Khinchin 1957). Equation (3.17) is then path-by-path, since source uncertainties can still be defined for individual paths. The 'game' equivalence classes still emerge directly, again leading to groupoid symmetry-breaking phase transitions.

It is further possible to explore the stochastic stability of the system of Eqs. (3.19) and (3.20) in much the same manner as was done previously. The basic stochastic differential equation is

$$dR_r = f(R_t)dt + \Omega h(R_t)dB_t \tag{3.21}$$

Here, $\Omega h(R)$ is the volatility term in the Brownian noise dB_t.

Application of the Ito Chain Rule to a stochastic stability function $Q(R)$ produces an analog to Eq. (3.12)—$< dQ_t >= 0$—as

$$f(R_t) \frac{d}{dR_t} Q(R_t) + \frac{\Omega^2 (h(R_t))^2 \frac{d^2}{dR_t^2} Q(R_t)}{2} = 0 \qquad (3.22)$$

Forcing stability in second order and with 'simple' volatility, so that $Q = R^2$ and $h(R) = R$, and assigning an 'exponential' relation as $dR/dt = f(R) = \beta - \alpha R$ determines the variance as

$$Var = \left(\frac{\beta}{\alpha - \Omega^2/2}\right)^2 - (\beta/\alpha)^2 \qquad (3.23)$$

This expression becomes explosively unstable as $\Omega \to \sqrt{2\alpha}$, implying that, for real world circumstances, it eventually becomes impossible to impose second-order stability under the burden of sufficient noise.

An Iterated Data Rate Theorem

Note that it is possible to derive an iterated Data Rate Theorem directly from the formalism of Eqs. (3.21) and (3.22), generalizing the results of Eqs. (3.10) and (3.14), by setting

$$f(R) = Rd^2F/dR^2 - M(\mathcal{H}) \qquad (3.24)$$

and solving for $M(\mathcal{H})$, where $F(R)$ is defined as in Eq. (3.18) and $M(\mathcal{H})$ as in Eq. (3.10).

Letting $Q(R) = R^2$ and $h(R) = R$—seeking second order stability under ordinary volatility—gives the stability condition as

$$M(\mathcal{H}) \geq \frac{R\Omega^2}{2} + Rd^2F/dR^2 \qquad (3.25)$$

As shown below, these matters can lead to unexpectedly complex dynamics.

3.3   Applications

Imposing Stability in Second Order: Aperiodic Stochastic Amplification

As Vazquez-Rodriguez et al. (2017) explain,

> [Stochastic resonance] was proposed as a possible explanation for the periodicity of the ice ages on Earth, and has been studied in Schmitt triggers, tunnel diodes and bidirectional ring lasers. Moreover, it was shown that stochastic noise plays a role in neuroautonomic regulation of the heart rate to generate complex dynamics like variability and scale invariance across a range of scales that are a hallmark of criticality.

Figure 3.2, adapted from Figure 1 of van der Groen et al. (2018), provides a snapshot introduction to periodic stochastic resonance. Groen et al. apply the phenomenon more generally to perceptual decision-making in humans, finding that adding an optimal level of neural noise to the visual cortex bilaterally enhanced decision-making for below-threshold stimuli, consistent with a form of stochastic resonance.

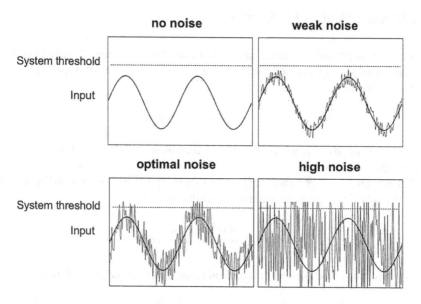

Fig. 3.2 Periodic stochastic resonance, as adapted from van der Groen et al. (2018). Stochastic resonance occurs when an optimal level of noise is added to a subthreshold signal. If the added noise is too weak, detection is not triggered, Too strong, and the underlying signal is washed out. Here, we impose stability in second order a priori, finding that increasing noise decreases signal detection threshold. In reality, sufficient noise makes such variance stabilization impossible, via mechanisms like Eq. (3.23)

The essential idea (Moss et al. 2004) is that when an optimal level of noise is added to a subthreshold signal, the rate of detection is significantly enhanced. Too little noise is ineffective, too much overwhelming—in our context via Eq. (3.23). Original studies were confined to periodic systems like Fig. 3.2. More recent work has extended the concept to generalized 'aperiodic' circumstance (e.g., Kang et al. 2020). As Kang et al. put it, for aperiodic stochastic resonance, instead of emphasizing frequency matching, the focus should be on what they call 'shape matching'. Here, we attempt to finesse some of this by explicitly *imposing* stability in second order—in variance—on the system of interest under Brownian noise.

We recover something much like an aperiodic stochastic amplification of signal detection threshold from our formalism, including the central role of criticality via Fisher zeros. As with the results of van der Groen et al., the effect is not simply confined to individual signal recognition, but acts across the full cognitive gestalt. We do, however, attempt to simplify the analysis by explicitly imposing stability in second order, i.e., in variance, by external mechanisms.

The second relation of Eq. (3.19),

$$dR/dt \approx dS/dR = Rd^2F(R)/dR^2$$

can be used to impose stochastic stability of a particular order on the system via the SDE

$$dR_t = (R_t d^2 F(R)/dR^2)dt + \Omega h(R_t)dB_t \tag{3.26}$$

Applying the Ito Chain Rule to the condition for stability in order Q, i.e., calculating $< dQ_t >= 0$ under volatility $h(R)$, gives

$$R\left(\frac{d^2}{dR^2}F(R)\right)\frac{d}{dR}Q(R) + \frac{\Omega^2 (h(R))^2 \frac{d^2}{dR^2}Q(R)}{2} = 0 \tag{3.27}$$

The general solution is

$$F(R) = \int -\frac{\Omega^2 (h(R))^2 \frac{d^2}{dR^2}Q(R)}{2R\frac{d}{dR}Q(R)} d[R, R] + C_1 R + C_2 \tag{3.28}$$

Restricting $Q(R) = R^2$ and $h(R) = R$—imposing stability in second order under 'simple' volatility in R—gives

$$F(R) = -\frac{\Omega^2 R^2}{4} + C_1 R + C_2 \tag{3.29}$$

It is interesting to compare this result with the Fisher-Eigen 'fitness function' adopted by Dunkel et al. (2004, Eq.(4)) in their study of stochastic resonance in nonlinear biological evolution models.

Assuming a continuum of possible 'levels' and approximating the probability sums by integrals in Eqs.(3.17) and (3.20), elementary calculation finds

$$F(R) \approx -\log[g(R)]g(R)$$

$$L \approx \exp[-H_0/g(R)] \tag{3.30}$$

Then

$$g = \frac{-F}{W(n, -F)}$$

$$g(R) = \frac{-(-\frac{\Omega^2 R^2}{4} + C_1 R + C_2)}{W(n, -(-\frac{\Omega^2 R^2}{4} + C_1 R + C_2))} \tag{3.31}$$

where $W(n, x)$ is the Lambert W-function of integer order n that satisfies

$$W(n, x) \exp[W(n, x)] = x.$$

The function is real-valued only for $n = 0, -1$ over the respective limited ranges $-\exp[-1] \leq x \leq \infty$ and $-\exp[-1] \leq x \leq 0$.

Thus a quite general condition for 'Q-order' stability—only real-valued expressions—in this model is

$$-F \geq -\exp[-1]$$

$$\frac{\Omega^2 R^2}{4} - C_1 R - C_2 \geq -\exp[-1] \tag{3.32}$$

where the second condition applies here to second-order stability in simple volatility.

Figure 3.3a examines the reaction rate $L(R)$ under these conditions, setting $n = 0$ and taking $F = -\Omega^2 R^2/4 - R/2 + 3/2$ and $H_0 = 1$ for two values of 'noise', $\Omega = 0, 1$. Figure 3.3b shows the stability condition for this model, illustrating the *decline* in critical value of R with increasing 'noise' Ω. This is not the 'inverted-U' of classic stochastic resonance, since we have imposed stability in second order—modulo the kinds of limitations implied by Eq. (3.23).

Under the condition of imposed stability in second order—either enforced by embedding regulators or by a simple limit on noise level—the system displays a highly punctuated form of aperiodic stochastic amplification. Here, we find the effect in full-bore cognition rate, not just periodic signal detection. To reiterate, this occurs *provided stability is imposed by external regulators or by a limit on noise amplitude*. That is, at a sufficiently high level of imposed noise Ω—Eq. (3.23)—the system breaks.

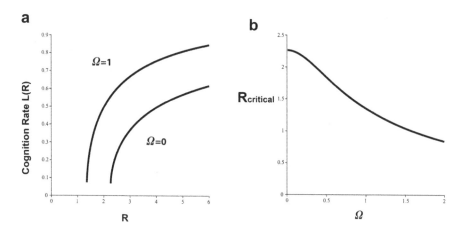

Fig. 3.3 Here, we impose stability in R^2 as $< dR_t^2 >= 0$ under simple volatility, i.e., $h(R) = R$. (**a**) A highly punctuated form of aperiodic stochastic amplification emerges for cognition rate under noise: higher noise level produces earlier punctuated onset of cognitive function, *provided stability in second order has been imposed*. At some level of applied noise Ω this becomes impossible, e.g., Eq. (3.23). (**b**) Systematic decline in critical value of R with increasing noise amplitude Ω under conditions of imposed second-order stability

It is 'easy' to replicate the calculation at the other end of the complexity spectrum, studying aperiodic stochastic amplification in a model with only two levels, having uncertainties $H \pm \delta$, with δ assumed small.

The partition function and cognition rate are then determined as

$$\exp[-F/g] = \exp[-(H+\delta)/g] + \exp[-(H-\delta)/g]$$

$$L(F) = \frac{1}{1 + \exp[2\delta/g]} \tag{3.33}$$

Expanding to second order, some manipulation finds the solution for g of most interest is

$$g \approx -\frac{F - H - \sqrt{-2\ln(2)\,\delta^2 + F^2 - 2FH + H^2}}{2\ln(2)} \tag{3.34}$$

Taking

$$F = \frac{-\Omega^2 R^2}{4} - \frac{R}{2} + \frac{11}{10} \tag{3.35}$$

and setting $H = 1$, $\delta = 0.1$ and $\Omega = 0, 1$ produces an analog to Fig. 3.3a in Fig. 3.4. Requiring that the expression under the square root sign in Eq. (3.34) be ≥ 0 generates a pattern roughly similar to Fig. 3.3b.

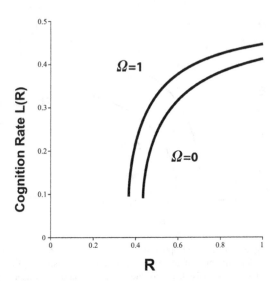

Fig. 3.4 The analog to Fig. 3.3a for a simple two-state system $H \pm \delta$, with δ assumed small. Again, Ω represents the added noise under the condition that stability in second order, i.e., $< dR_t^2 >= 0$, has been imposed for simple volatility, $h(R) = R$. Requiring the expression under the square root in Eq. (3.34) to be ≥ 0 produces a pattern analogous to Fig. 3.3b

General results at intermediate scales for dependence of cognition rate on Ω under conditions of enforced second-order stability can be obtained at the expense of some considerable calculation. An essential feature of the argument is that, from Eq. (3.29)—under imposed second-order stability—F is concave: $d^2 F / dR^2 = -\Omega^2 / 2 \leq 0$.

An acute observer will have perhaps noted that this result suggests a means for generalization to 'colored' noise via the expression

$$d^2 F / dR^2 = -\frac{1}{2} d[Y_t, Y_t]/dt \tag{3.36}$$

where Y_t is an appropriate SDE and $[Y_t, Y_t]$ its quadratic variation (Protter 2005). Extension to Levy jump processes should be direct. See Kang et al. (2020) and Wallace (2016) for roughly analogous discussions.

Imposing Other Forms of Stability

The dynamic behavior of stochastic systems, in particular the stability of stochastic differential equations, is, if anything, vastly overrich (e.g., Appleby et al. 2008; Protter 2005).

Carrying through Eq. (3.28) for R^n—calculating $< dR_t^n >= 0$—again taking $h(R) = R$, leads to the relations

$$F(R) = -\frac{\Omega^2(n-1)R^2}{4} + C_1 R + C_2$$

$$d^2 F/dR^2 = -\frac{\Omega^2(n-1)R^2}{2} \tag{3.37}$$

If $n > 1$, the results will be analogous to the development above. If $n < 1$, the sign of the relations changes, and the world is different, becoming analogous to the case in which the behavior of interest depends, not on R^n, $n > 1$, but on $\log(R)$, recalling that logarithmic biophysical laws are characteristic of many sensory systems (e.g., Adler et al. 2014). Then, we calculate conditions under which $< d\log(R_t) >= 0$, again taking $h(R) = R$, so that

$$F = \frac{\Omega^2 R^2}{4} + C_1 R + C_2$$

$$d^2 F/dR^2 = \frac{\Omega^2}{2} \tag{3.38}$$

Details, making the integral approximation to Eq. (3.17) that leads to Eqs. (3.30) and (3.31), are relatively straightforward to work out, noting that the sign of the term in Ω is *positive* instead of negative as in the calculations leading to Fig. 3.3. Here, we set $C_1 = -1$, $C_2 = 1$.

Imposing stability as $< d\log[R_t] >= 0$, Fig. 3.5a shows the cognition rate as a function of R for two values of noise Ω. $\Omega = 0$ matches Figs. 3.3a and 3.4, while $\Omega = 1$ generates a Yerkes-Dodson inverted-U signal transduction form (Diamond et al. 2007). Figure 3.5b shows cognition rate as a joint function of Ω and R, displaying a full collapse at sufficient noise level. Figure 3.5c shows the details of that collapse.

We reiterate that, for the integral model leading to Eq. (3.28), based on the Lambert W-function of orders 0 or -1, a necessary condition for stability is that $-F \geq -\exp[-1]$, placing noise-driven constraints on Q-order stability, particularly for the $\log(R_t)$ model and the R_t^n model with $n < 1$.

Regulating Inherent Instability

Inherently unstable cognitive systems must be heavily regulated, becoming subject to constraints of the Data Rate Theorem, as indicated by the developments of Eqs. (3.24) and (3.25). The dynamics of Figs. 3.3 and 3.5 are, in consequence, greatly altered.

The essential point is that $Rd^2 F/dR^2 \rightarrow Rd^2 F/dR^2 - M(R)$ in Eq. (3.26) so that Eq. (3.28) becomes

Fig. 3.5 Here, stability has been imposed as $< d \log[R_t)] >= 0$ under volatility $h(R) = R$. (**a**) $L(R)$ as a function of R for two values of Ω. Sufficient noise produces a Yerkes-Dodson inverted-U (Diamond et al. 2007). (**b**) The three dimensional model of $L(R)$ vs. Ω and R, showing the collapse at sufficiently large noise. (**c**) Details of cognitive collapse at sufficient noise

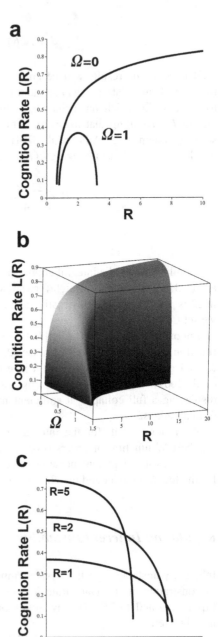

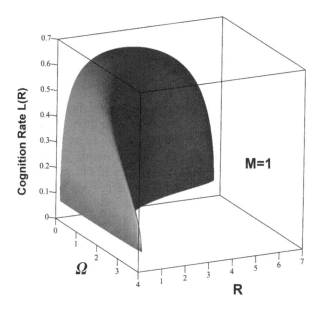

Fig. 3.6 Replication of the calculation leading to Fig. 3.5b, based on the relation $< d \log(R_t) > = 0$, taking $M = 1$ in Eq. (3.39). Logarithmic biophysical laws are characteristic of sensory systems

$$F(R) =$$

$$\iint \frac{-\Omega^2 (h(R))^2 \frac{d^2}{dR^2} Q(R) + 2 M(R) \frac{d}{dR} Q(R)}{2 R \frac{d}{dR} Q(R)} \, dR \, dR$$

$$+ C_1 R + C_2 \qquad (3.39)$$

For M constant, carrying through the calculations leading to Figs. 3.3 and 3.5—assuming Eq. (3.30)—finds that, taking $Q(R) = R^2$, the cognition rate pattern now resembles Fig. 3.5a—stochastic amplification disappears—and for $Q(R) = \log(R)$, the cognition rate becomes a Yerkes-Dodson inverted-U even for zero Ω that completely collapses at larger values.

Figure 3.6 replicates the calculation leading to Fig. 3.5b, which was equivalent to setting $M = 0$, but takes $M = 1$.

Again, for $Q(R) = R^2$, a particularly interesting example emerges if $M(R) = M R$ under simple volatility, i.e., $h(R) = R$.

Detailed calculations are left as an exercise.

Some Implications of These Examples

The general approach of this study may contribute toward explanation of the wide spectrum of dynamic patterning found across sensory systems. As Adler et al. (2014) describe,

Two central biophysical laws describe sensory responses to input signals. One is a logarithmic relationship between input and output, and the other is a power law relationship. These laws are sometimes called the Weber-Fechner law and the Stevens power law, respectively. The two laws are found in a wide variety of human sensory systems including hearing, vision, taste, and weight perception; they also occur in the responses of cells to stimuli. However the mechanistic origin of these laws is not fully understood...

The work here suggests that sensory dynamics—and, indeed, cognitive dynamics in general—may depend on the details of embedding Data Rate Theorem-related regulatory stabilization under the burdens of noise. Only under particular circumstances—here, $n > 1$ in Eq. (3.37) (given simple volatility)—will global cognitive phenomena akin to aperiodic stochastic resonance occur, with 'signal transduction' and other patterns similar to Fig. 3.5 perhaps more common, particularly in the case of inherent instability.

3.4 Discussion

The Rate Distortion Theorem arguments above lead to very general forms of the Data Rate Theorem, usually derived as an extension of the Bode Integral Theorem (e.g., Nair et al. 2007, and references therein). The approach places control theory and information theory within the same milieu, encompassing many of the cognitive phenomena so uniquely characterizing the living state (e.g., Maturana and Varela 1980).

Specifically, these arguments imply that there cannot be cognition—held within the confines of information theory—without a parallel regulation, held within the confines of control theory. Both theories are constrained by powerful—and apparently closely related—asymptotic limit theorems.

The generalization of Eq. (3.17) et seq. extends the approach to a multiplicity of different 'games' by invoking equivalence classes of control path sequences, generating an iterated model. The partition function—the denominator of Eq. (3.17)—allows definition of iterated free energy, its associated Legendre transform entropy, and an iterated Onsager approximation in the gradient of that entropy analog driving system dynamics.

Wallace (2021) explores extension of the 'entropy' construct, and of the Onsager gradient model defining system dynamics, to higher orders, focusing on the use of formal power series ring structures. A central outcome of that work is that the 'spectra' of the formal power series coefficients will characterize essential features of the systems under study, and indeed may have their own dynamics under changing patterns of threat and affordance. A simple example of such extension is given in the Mathematical Appendix.

Stochastic stability analysis of the simple linear Onsager model is direct. However, an extended perspective focuses on groupoid symmetry-breaking phase transitions driven by Fisher Zero analogs.

In addition, the model uncovers an aperiodic stochastic amplification, provided that '$n > 1$' stability from Eq. (3.37) has been imposed by external regulation. If this is not possible—if regulation fails or noise becomes overwhelming, for example via mechanisms like Eq. (3.23)—then cognition itself fails.

Indeed, extension of the approach to different forms of imposed stabilization—different $Q(R)$ in Eq. (3.27)—seems to provide insight into the varieties of input-output relations in sensory systems, modulo the almost infinite varieties of stability modes and conditions afflicting stochastic differential equations (Appleby et al. 2008).

We have outlined a novel approach to building new statistical tools for data analysis in cognitive systems, focusing on groupoid symmetry-breaking phase transitions characterized by Fisher Zero analogs, and by the spectrum of stochastic stabilization and de-stabilization. The methodology suggests the possibility of importing tools from cutting edge theory into cutting edge empirical study.

3.5 Mathematical Appendix

Imposing Higher Order Stability

Following Wallace (2021), a relatively simple extension of theory can be made by redefining an 'entropy' Eq. (3.19) as

$$S_1 = -F(R) + RdF/dR + \epsilon R^2 d^2 F/dR^2 \qquad (3.40)$$

Retaining the first-order Onsager approximation,

$$dR/dt \approx dS_1/dR =$$
$$Rd^2F/dR^2(1 + 2\epsilon) + \epsilon R^2 d^3 F/dR^3 \qquad (3.41)$$

The stochastic differential equation of interest is then

$$dR_t = \left(Rd^2F/dR^2(1 + 2\epsilon) + \epsilon R^2 d^3 F/dR^3 \right) dt +$$
$$\Omega V(R_t)dB_t \qquad (3.42)$$

Imposing the stability condition $< dQ_t >= 0$ via the Ito Chain Rule—analogous to Eq. (3.28)—leads to the general expression

$$F(R) =$$
$$\int -\frac{1}{2\epsilon} R^{-\frac{1+2\epsilon}{\epsilon}}$$

$$\left(\int \frac{(V(R))^2 \left(\frac{d^2}{dR^2} Q(R)\right) R^{\epsilon-1}}{\frac{d}{dR} Q(R)} \, dR \Omega^2 - 2 C_1 \epsilon\right) d[R, R]$$

$$+ C_2 R + C_3 \qquad (3.43)$$

Taking $Q(R) = R^n$ and $V(R) = R$ leads to the relation

$$F(R) =$$

$$-\frac{C_1 \epsilon}{\epsilon + 1} R^{2 - \frac{1+2\epsilon}{\epsilon}} \left(2 - \frac{1 + 2\epsilon}{\epsilon}\right)^{-1} - \frac{\Omega^2 (n-1) R^2}{4 + 8\epsilon} +$$

$$C_2 R + C_3 \qquad (3.44)$$

analogous to Eq. (3.37).

Expanding the Onsager entropy gradient model in Eq. (3.41) can also be done. For example, in some contrast to Eq. (3.31) in Wallace (2021), we make a *dimensionally-consistent* expansion

$$dR/dt \approx dS_1/dR + \delta R d^2 S_1/dR^2 \qquad (3.45)$$

in the SDE of Eq. (3.42) that produces a somewhat more complicated version of Eq. (3.43).

The $Q(R) = R^n$ model under simple volatility, i.e. $V(R) = R$, gives

$$F(R) =$$

$$\frac{C_1 R^{2 - \frac{2\epsilon+1}{\epsilon}}}{\left(1 - \frac{2\epsilon+1}{\epsilon}\right) \left(2 - \frac{2\epsilon+1}{\epsilon}\right)} +$$

$$\frac{C_2 R^{2 - \frac{\delta+1}{\delta}}}{\left(1 - \frac{\delta+1}{\delta}\right) \left(2 - \frac{\delta+1}{\delta}\right)} -$$

$$\frac{\Omega^2 (n-1) R^2}{2 ((4\delta + 4) \epsilon + 2\delta + 2)} +$$

$$C_3 R + C_4 \qquad (3.46)$$

Imposing logarithmic stability, i.e., setting $Q(R) = \log[R]$, and taking $V(R) = R$, gives

$$F(R) =$$

$$\frac{\Omega^2 R^2}{2 (4\delta + 4) \epsilon + 4\delta + 4} +$$

$$C_1 R^{-\epsilon^{-1}} \left(1 - \frac{2\epsilon + 1}{\epsilon}\right)^{-1} \left(2 - \frac{2\epsilon + 1}{\epsilon}\right)^{-1} +$$

$$C_2 R^{\frac{\delta - 1}{\delta}} \left(1 - \frac{\delta + 1}{\delta}\right)^{-1} \left(2 - \frac{\delta + 1}{\delta}\right)^{-1} +$$

$$C_3 R + C_4 \tag{3.47}$$

Note the similarities and differences in the Ω terms between Eqs. (3.46) and (3.47).

Here, ϵ and δ constitute the 'spectra' of the series expansion.

It is interesting to reconsider the model of Fig. 3.5 from this perspective, taking $\delta = \epsilon = -1/2$. Here, again $Q(R) = \log(R)$, $V(R) = R$. The condition $< dQ_t >= 0$ in the SDE calculation leads to the determination of F, g and L as

$$\frac{R^2}{4} d^4 F / d R^4 - \frac{\Omega^2}{2} = 0$$

$$F(R) =$$

$$\frac{C_2 R^2}{2} + \frac{3 \Omega^2 R^2}{2} + \frac{C_1 R^3}{6} -$$

$$\Omega^2 R^2 \ln(R) + C_3 R + C_4$$

$$g(R) = \frac{-F}{W(0, -F)}, \quad L = \exp[-1/g(R)] \tag{3.48}$$

Figure 3.7 shows the cognition rate $L(R)$, taking $C_1 = 1$, $C_2 = C_3 = -1$, $C_4 = 1/2$, and setting $\Omega = 0, 6/7, 5/2$. At zero stochastic burden—$\Omega = 0$—cognition rate shows a characteristic signal transduction inverted-U Yerkes-Dodson pattern. As Ω increases, the inverted-U collapses in on itself. At a critical noise level, however, stochastic burden drives the system into a 'panic mode' breakdown, most characteristic of the collapse of institutional cognition during conflict (Wallace 2022c).

The methodology, even in this relatively simple form, can evidently encompass a vast range of dynamic cognitive behaviors.

Seeking Higher Order Stability

A different approach seeks stability under the influence of noise, rather than studying dynamics on the assumption of stability, as in the previous section. Here, We again take second-order approximations in the definition of S and in the Onsager approximation, and again set $\delta = \epsilon = -1/2$. The assumption, however, is that the ability to provide channel capacity R is itself dynamic, setting $dR/dt = \beta - \alpha R$,

Fig. 3.7 Cognition rate vs. R for the model of Eq. (3.48). Here, $\epsilon = \delta = -1/2$, with $C_1 = 1$, $C_2 = C_3 = -1$, $C_4 = 1/2$. As stochastic burden Ω rises, the system first sees the collapse of a Yerkes-Dodson signal transduction inverted-U, followed by punctuated onset of a 'panic mode' response. This behavior is particularly characteristic of institutional failure during conflict (Wallace 2022c)

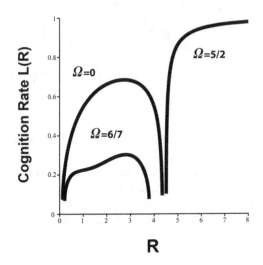

leading to an 'exponential' model under imposed noise Ω. Here, we obtain the basic relation in F as

$$\frac{R^3 \frac{d^4}{dR^4} F(R)}{4} = \beta - \alpha R$$

$$F(R) =$$

$$\frac{C_2 R^2}{2} - 3 R^2 \alpha + 2 \ln(R) R\beta - 2 R\beta + \frac{C_1 R^3}{6} +$$

$$2 \alpha R^2 \ln(R) + C_3 R + C_4 \qquad (3.49)$$

with g and L defined as in Eq. (3.46).

The basic stochastic differential equation, however, is now

$$dR_t = (\beta - \alpha R_t)dt + \Omega R_t dB_t \qquad (3.50)$$

where dB_t is taken as Brownian white noise.

An important point is that R^2 is now subject to noise constraint, so that the condition for stability in variance of R is now found by the Ito Chain Rule as in Eq. (3.23), i.e.,

$$Var = \left(\frac{\beta}{\alpha - \Omega^2/2}\right)^2 - \left(\frac{\beta}{\alpha}\right)^2$$

Setting $\alpha = 1$, $C_1 = 1$, $C_2 = C_3 = -1$, $C_4 = 1/2$, and letting β vary produces a Yerkes-Dodson signal transduction for the cognition rate $L(R)$ in Fig. 3.8a.

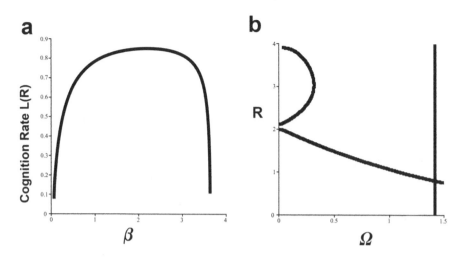

Fig. 3.8 (a) Inverted-U cognition rate under increasing channel capacity R, taking $\alpha = 1$, $C_1 = 1$, $C_2 = C_3 = -1$, $C_4 = 1/2$. Note the peak at $\beta \approx 2$. (b) Solution of $< dL_t >= 0$ under increasing stochastic burden Ω. The vertical line is the limit $\Omega = \sqrt{2}$. Note the possible onset of bifurcation instability in the cognition rate at low levels of Ω

Figure 3.8b imposes stochastic burden Ω on the cognition rate $L(R)$ via the Ito Chain Rule, setting $\beta = 2$, near the peak of Fig. 3.8a, and numerically solving the relation $< dL_t >= 0$. The vertical line on the right is the stability limit for variance in R, here $\Omega = \sqrt{2}$ since $\alpha = 1$. At $\Omega \rightarrow \approx 0.31$—much less than the variance limit $\sqrt{2}$—the cognition rate risks onset of a bifurcation instability.

A slightly different approach uses an 'Arrhenius' model for which $R(t) = \beta \exp[-\alpha/t] \rightarrow \beta$, so that, following Eq. (3.49),

$$\frac{1}{4}R^3 d^4 F(R)dR^4 = dR/dt = \frac{R}{\alpha}(\log[R/\beta])^2 \tag{3.51}$$

Proper selection of boundary conditions again leads to inverted-U transduction forms for $L(R)$ as depending on β for fixed α, and to an analogous—but distinctly different—pattern of stochastic dynamics for $L(R)$ under the Ito Chain Rule.

Extending the Model

Other, far more complicated, forms of Eqs. (3.40) and (3.45) are, of course, also both possible and likely. For example, *dimensionally-consistent and without cross-terms*, formal series expansions can be taken as

$$S \equiv -F(R) + R \, dF/dR + \sum_{n=2}^{\infty} \epsilon_n R^n d^n F/dR^n$$

$$\partial R/\partial t = \mu \, dS/dR + \sum_{n=2}^{\infty} \mu_n R^{n-1} d^n S/dR^n \tag{3.52}$$

It is worth placing such expansions explicitly within the domain described by Jackson et al. (2017), who write

[T]he Legendre transform can be viewed as mapping the coefficients of one formal power series into the coefficients of another formal power series. Here, the term 'formal' does not express 'mathematically nonrigorous', as it often does in the physics literature. Instead, the term 'formal power series' is here a technical mathematical term, meaning a power series in indeterminates. Formal power series are not functions. A priori, formal power series merely obey the axioms of a ring and questions of convergence do not arise.

For our context, in some contrast to bare-bones physics developments, these matters—the ϵ and μ spectra and possible series crossterms—involve fundamentally empirical questions associated with a statistical model, much as in the fitting of ordinary regression models to data sets.

References

Adler, M., A. Mayo, and U. Alon, 2014. Logarithmic and power law input-output relations in sensory systems with fold-change detection. PLOS Computational Biology https://doi.org/10.1371/journal.pcbi.1003781.

Appleby, J., X. Mao, and A. Rodkina, 2008. Stabilization and destabilization of nonlinear differential equations by noise. IEEE Transactions on Automatic Control 53:126–132.

Atlan H., and I. Cohen, 1998. Immune information, self-organization, and meaning. International Immunology 10:711–718.

Cover, T., and J. Thomas, 2006. *Elements of Information Theory*, 2nd ed. New York: Wiley.

de Groot, S., and P. Mazur, 1984. *Nonequilibrium Thermodynamics*. New York: Dover.

Diamond, D., A. Campbell, C. Park, J. Halonen, and P. Zoladz, 2007. *The Temporal Dynamics Model of Emotional Memory Processing Neural Plasticity*. https://doi.org/10.1155/2007/60803.

Dolan, B., W. Janke, D. Johnston, and M. Stathakopoulos, 2001. Thin Fisher zeros. Journal of Physics A 34:6211–6223.

Dunkel, J., S. Hilbert, L. Schimansky-Geier, and P. Hanggi, 2004. Stochastic resonance in biological nonlinear evolution models. Physical Review E 69:056118.

Effros, M., P. Chou, and R. Gray, 1994. Variable-rate source coding theorems for stationary nonergodic sources. IEEE Transactions on Information Theory 40:1920–1925.

Einstein, A., 1905/1956. *Investigations on the Theory of the Brownian Motion*. New York: Dover Publications.

Feynman, R., 2000. *Lectures on Computation*. New York: Westview Press.

Fisher, M., 1965. *Lectures in Theoretical Physics*, Vol. 7. Boulder: University of Colorado Press.

Jackson, D., A. Kempf, and A. Morales, 2017. A robust generalization of the Legendre transform for QFT. Journal of Physics A 50:225201.

Kang, Y., R. Liu, and X. Mao, 2020. Aperiodic stochastic resonance in neural information processing with Gaussian colored noise. Cognitive Neurodynamics https://doi.org/10.1007/s11571-020-09632-3.

Khinchin, A., 1957. *Mathematical Foundations of Information Theory*. New York: Dover.

Laidler, K., 1987. *Chemical Kinetics*, 3rd ed. New York: Harper and Row.

Maturana, H., and F. Varela, 1980. *Autopoiesis and Cognition: The Realization of the Living*. Boston: Reidel.

Moss, F., L. Ward, and W. Sannita, 2004. Stochastic resonance and sensory information processing: a tutorial and review of application. Clinical Neurophysiology 115:267–281.

Nair, G., F. Fagnani, S. Zampieri, and R. Evans, 2007. Feedback control under data rate constraints: an overview. Proceedings of the IEEE 95, 108138.

Protter, P., 2005. *Stochastic Integration and Dierential Equations*, 2nd ed. New York: Springer.

Ruelle, D., 1964. Cluster property of the correlation functions of classical gases. Reviews of Modern Physics, April:580–584.

Shields, P., D. Neuhoff, L. Davisson, and F. Ledrappier, 1978. The Distortion-Rate function for nonergodic sources. The Annals of Probability 6:138–143.

van der Groen, O., M. Tang, N. Wenderoth, and J. Mattingley, 2018. Stochastic resonance enhances the rate of evidence accumulation during combined brain stimulation and perceptual decision-making. PLOS Computational Biology 14:e1006301. https://doi.org/10.1371/journal.pcbi.1006301.

Vazquez-Rodriguez, B., A. Avena-Koenigsberger, O. Sporns, A. Griffa, P. Hagmann, and H. Larralde, 2017. Stochastic resonance at criticality in a network model of the human brain. Scientific Reports 7:13020 https://doi.org/10.1038/s41598-017-13400-5.

Wallace, R., 2005. *Consciousness: A Mathematical Treatment of the Global Neuronal Workspace Model*. New York: Springer.

Wallace, R., 2012. Consciousness, crosstalk, and the mereological fallacy: an evolutionary perspective. Physics of Life Reviews 9:426–453.

Wallace, R., 2016. Subtle noise structures as control signals in high-order biocognition. Physics Letters A 380:726–729.

Wallace, R., 2020. On the variety of cognitive temperatures and their symmetry-breaking dynamics. Acta Biotheoretica. https://doi.org/10.1007/s10441-019-09375-8.

Wallace, R., 2021. Toward a formal theory of embodied cognition. BioSystems 202:104356.

Wallace, R., 2022a. *Consciousness, Cognition and Crosstalk: The Evolutionary Exaptation of Nonergodic Groupoid Symmetry-Breaking*. New York: Springer.

Wallace, R., 2022b. Formal perspectives on shared interbrain activity in social communication. Cognitive Neurodynamics. https://doi.org/10.1007/s11571-022-09811-4.

Wallace, R., 2022c. *Deception and Delay in Organized Conflict: Essays on the Mathematical Theory of Maskirovka*. New York: Springer.

Weinstein, A., 1996. Groupoids: unifying internal and external symmetry. Notices of the American Mathematical Association 43:744–752.

Chapter 4
Punctuated Regulation as an Evolutionary Mechanism

We...suggest [for evolutionary process]... that ancient regulatory programs that have been more or less modified continue to be utilized. As development progresses from stage to stage, progressively less ancient and phylogenetically more restricted genomic regulatory patterns would come into play.
— Britten and Davidson (1971)

Since the morphological features of an animal are the product of its developmental process, and since the developmental process in each animal is encoded in its species-specific regulatory genome, then change in animal form during evolution is the consequence of change in genomic regulatory programs for development.
— Davidson (2006)

4.1 Introduction

Casanova and Konkel (2020) study specific genetic mechanisms supposed to underlie Eldredge/Gould evolutionary punctuated equilibrium (Gould 2002, Ch.9), via a two-part hypothesis involving stasis through highly-conserved developmental mechanisms that are challenged by accumulation of regulatory elements and recombination within these same genes, leading to punctuated bursts of morphological divergence and speciation across metazoa.

Here, adapting the formalisms of previous chapters, we explore more general phenomena associated with heritage and regulation that imply similar dynamics, but are not constrained to purely genetic explanation, consonant with the spirit of the Extended Evolutionary Synthesis.

4.2 Fisher Zeros Reconsidered

We review the iterated argument of Chap. 3, focusing again on ensembles of possible regulatory processes:

1. Control dynamics—the set of 'Data Rate Theorems' (DRT) available to a system—are context-dependent. Different 'road conditions' require different analogs to, or forms of, the DRT. More specifically, the transmission of control messages as represented in the representation of 'average distortion' between what is ordered and what is observed in a control process (e.g., Eq. (3.11)) defines equivalence classes of possible control sequences $X \equiv \{X_t, X_{t+1}, X_{t+2}, ...\}$. Again, the example of driving along a particular stretch of road slowly on a dry, sunny morning is a different 'game' from driving the same stretch at high speed during a midnight snowstorm.

2. The set of all possible such 'games' can be characterized in terms of equivalence classes G_j of the control path sequences X. Each class is then associated with a particular cognitive 'game' represented by a 'dual' information source having uncertainty

$$HG_j \equiv H_j \tag{4.1}$$

that will vary across the complexity of the 'games' being played. Again, dry, sunny morning vs snowstorm night driving.

A 'dual' information source arises from the necessity of making behavioral choices in such 'game-playing'.

To reiterate, choice reduces uncertainty, and the reduction of uncertainty necessarily implies existence of an information source dual to the cognitive process. Different 'games' are associated with different 'languages' of play involving such choice, all having different dual information sources.

3. Here, the Rate Distortion Function $R(D)$, associated with a particular average distortion D for a particular 'vehicle', is imposed on system dynamics via temperature analog $g(R)$ in an iterated development through a Boltzmann-like pseudoprobability

$$P_j = \frac{\exp[-H_i/g(R)]}{\sum_j \exp[-H_i/g(R)]} \tag{4.2}$$

where, unlike 'ordinary' statistical thermodynamics, the 'temperature' $g(R)$ must be calculated from first principles.

4. The denominator of Eq. (4.2) is taken as representing an analog to the standard statistical mechanical partition function, and used to derive an iterated 'free energy' analog F as

$$\exp[-F/g(R)] = \sum_k \exp[-H_k/g(R)] \equiv A(g(R))$$

$$F(R) = -\log[A(g(R))]g(R) \tag{4.3}$$

5. It is now possible to define an 'entropy' through a classic Legendre transform on F, and to impose system dynamics through the first-order Onsager approxi-

mation of nonequilibrium thermodynamics, defining 'thermodynamic forces' in terms of the gradient of that 'entropy' (de Groot and Mazur 1984):

$$S(R) \equiv -F(R) + R dF/dR$$

$$\partial R/\partial t \approx dS/dR = R d^2 F/dR^2 = f(R)$$

$$F(R) = \int \frac{f(R)}{R} d[R, R] + C_1 R + C_2$$

$$g(R) = \frac{-F(R)}{Root\,Of\,(\exp[Z] - A(-F/Z))} \tag{4.4}$$

These relations are not simple:

- $f(R)$ represents a 'friction' inherent to any control system, characterizing the rate it can react to an action signal, for example, via an 'exponential model' $dR/dt = f(R) = \beta - \alpha R$, $R(t) \to \beta/\alpha$.
- As discussed previously, the *RootOf* construction generalizes the Lambert W-function $W(n, x)$ of order n, seen by carrying through a calculation setting $A(g(R)) = g(R)$, so that

$$F = -\log[g]g$$

$$g = \frac{-F}{W(n, -F)} \tag{4.5}$$

where $W(n, x)$ is real-valued only for $n = 0, -1$ and only for $x \geq -\exp[-1]$.

- Most significantly for this analysis, however, since the 'RootOf' construction may have complex number solutions, the temperature analog $g(R)$ directly imposes 'Fisher Zeros' closely analogous to those associated with 'ordinary' phase transition in physical systems (Dolan et al. 2001; Fisher 1965; Ruelle 1964, Sec. 5).
- The set of equivalence classes $\{G_j\}$ defines a groupoid (Weinstein 1996), and the Fisher Zeros represent groupoid symmetry-breaking for cognitive phenomena that is analogous to, but markedly different from, the group symmetry-breaking associated with phase transition in physical processes.
- As a result, groupoid symmetry-breaking phase transitions represent a significant extension of the basic control theory Data Rate Theorem (Nair et al. 2007).

In a very fundamental sense, we have directly—and greatly—generalized the inferences of Casanova and Konkel (2020), moving beyond the strictly genetic to all possible modalities of the inheritance/regulation dyad.

4.3 Extinction I: Simple Noise-Induced Transitions

A large literature explores noise-induced transitions in dynamic systems (e.g., Van den Broeck et al. 1994, 1997; Horsthemeke and Lefever 2006; Tian et al. 2017), and we adapt something of that to the model of Eqs. (4.4) and (4.5), taking $f(R) = \beta - \alpha R$ in Eq. (4.4).

Then

$$F(R) = R\beta \log[R] - R\beta - \frac{R^2\alpha}{2} + C_1R + C_2 \tag{4.6}$$

where C_1 and C_2 are appropriate boundary conditions.

Taking the Lambert W-function of order $n = 0$, stability in the temperature analog $g(R)$ is, in part, determined by the condition $-F \geq -\exp[-1]$.

Figure 4.1 explores system dynamics by calculating g at the nonequilibrium steady state $R \rightarrow \beta/\alpha$, taking $\alpha = 1$, $C_1 = -1$, $C_2 = 1$.

The basic pattern is one of Yerkes-Dodson 'inverted-U' signal transduction.

From the second expression of Eq. (4.4) we construct a basic Stochastic Differential Equation for the dynamics of this system under noise as

$$dR_t = (\beta - \alpha R_t)dt + \sigma R_t dB_t \tag{4.7}$$

where the second term represents 'ordinary' volatility in standard Brownian noise.

Application of the Ito Chain Rule (Protter 2005) to $<dR_t^2> = 0$ via Eq. (4.7) allows calculation of variance as

Fig. 4.1 Setting $R \rightarrow \beta/\alpha$ in the calculation of the temperature analog $g(R)$ generates a Yerkes-Dodson inverted-U signal transduction pattern. Here, $\alpha = 1$, $C_1 = -1$, $C_2 = 1$

$g(\beta)$

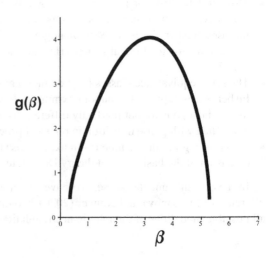

β

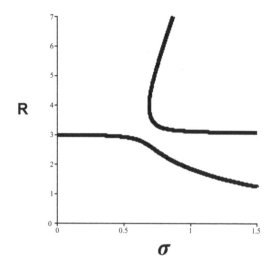

Fig. 4.2 Solution set $\{R, \sigma\}$ for the relation $< dg_t >= 0$ from Fig. 4.1. Here, $\beta = 3$, $\alpha = 1$, $C_1 = -1$, $C_2 = 1$. The solution set shows a phase transition bifurcation at $\sigma \approx 0.69$, far lower than the limit $\sigma < \sqrt{2\alpha}$ required for stability in variance

$$Var = \left(\frac{\beta}{\alpha - \sigma^2/2}\right)^2 - \left(\frac{\beta}{\alpha}\right)^2 \tag{4.8}$$

so that the system becomes explosively unstable in second order as $\sigma \to \sqrt{2\alpha}$.

Fixing β at the maximum of Fig. 4.1—$\beta = 3$—the Ito Chain Rule also permits determination of the nonequilibrium steady state condition $< dg_t >= 0$ in a similar manner. Numerical calculation determines the solution set $\{R, \sigma\}$ of that equation in Fig. 4.2. A bifurcation emerges at $\sigma \approx 0.69$, much less than the variance stability condition $\sigma < \sqrt{2}$ for this system. Such bifurcations, it may perhaps be asserted, constitute serious selection pressures that challenge population survival.

In the Mathematical Appendix we revisit this calculation using the extended definitions of 'entropy' and the entropy gradient relation of Eqs. (3.40) and (3.47).

4.4   Extinction II: More Complicated Noise-Induced Transitions

A different perspective emerges if the system is inherently unstable in the sense of the previous chapter, analogous to Eq. (3.12), then, for simple volatility, Eq. (3.26) becomes

$$dR_t = \left(Rd^2F/dR^2 - M(\mathscr{H})\right)dt + \sigma R_t dB_t \tag{4.9}$$

where $M(\mathcal{H})$ is the control free energy needed to impose stability in the particular order chosen, and dB_t is Brownian noise. We impose stability in second order, seeking solutions via the Ito Chain Rule for the relation $< dR_t^2 >= 0$.

Direct calculation finds

$$F(R) = M \ln(R)R - MR - \frac{\sigma^2 R^2}{4} + C_1 R + C_2 \tag{4.10}$$

where we assume a continuous system so that, following the arguments of Eq. (3.30), we can define a temperature analog g and related reaction rate L as

$$g = \frac{-F}{W(0, -F)}$$

$$L = \exp[-H_0/g] \tag{4.11}$$

Again, $W(n, x)$ is the Lambert W-function of order n.

Figure 4.3 shows $F(\sigma, R)$ for $M = 0, 1, C_1 = -1, C_2 = 1$. The plane is the stability limit $-F \geq -\exp[-1]$, since, for differing M, $g(\sigma, R)$ can be real-valued only under that condition, setting severe—but different—limits on population survival under noise.

Figure 4.4 shows the corresponding reaction rates L, taking $H_0 = 1$. For $M = 1$ the system displays 'inverted-U' signal transduction at low values of σ, essentially a kind of Yerkes-Dodson effect. These dynamics are in distinct contrast to the

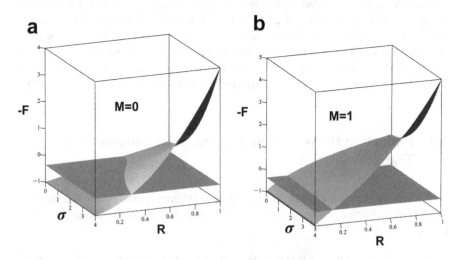

Fig. 4.3 $F(\sigma, R)$ according to Eq. (4.10), assuming an inherently unstable system requiring control free energy at rates. (a) $M = 0$, (b) $M = 1$. For both, $C_1 = -1$, $C_2 = 1$. The real-value condition on the Lambert W-function defining the temperature analog g for a continuous system, that $-F \geq -\exp[-1]$, the plane in both figures, places severe, but different, constraints on population survival under noise

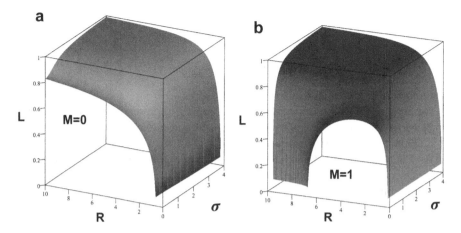

Fig. 4.4 'Reaction rate' $L(\sigma, R)$ according to the second part of Eq. (4.11). (**a**). The $M = 0$ system shows simple dynamics across both σ and R. (**b**). At $M = 1$ the system displays a Yerkes-Dodson signal transduction pattern at low σ which disappears at higher values. This can be considered a kind of analog to stochastic resonance in regulated systems

results of Fig. 3.5, which considers a similar system *without* imposition of control information at the rate M.

4.5 Extinction III: Environmental Shadow Price

It is also possible to reconsider the selection pressure/shadow price argument of the previous chapters, but now in terms of a multicomponent inheritance/regulation system and the relations

$$\partial R_i / \partial t = dS_i / dR_i = f_i(R_i) \tag{4.12}$$

where the entropy analogs S_i emerge from Eq. (4.4).

Again, the complete system consists of $i = 1, 2, ..., m$ cooperating units that must individually supplied with bandwidth at rates R_i under the overall constraints of resources and time as $\sum_i R_i = R$, $\sum_i T_i = T$. We then optimize overall distortion under these constraints, assuming $dR_i/dt = f(R_i) \to 0$.

As discussed above, while there are general approaches to optimization (e.g., Nocedal and Wright 2006), we again construct a general Lagrangian optimization—minimization—across subcomponent distortions D_i as

$$\mathscr{L} = \sum_i D_i + \lambda \left(R - \sum_i R_i \right) + \mu \left(T - \sum_i T_i \right)$$

$$\partial \mathscr{L}/\partial R = \lambda$$

$$\partial D_i/\partial R_i = \lambda$$

$$\partial D_i/\partial T_i = \mu \qquad (4.13)$$

remembering in particular that $\partial R_i/\partial T_i = f_i(R_i)$.

Recall that, for simple economic models, λ and μ represent the 'shadow prices' imposed on system dynamics by environmental constraints, in a large sense (Jin et al. 2008; Robinson 1993).

As has been discussed, elementary Lagrangian optimization, while admittedly limited, serves to outline the basic mechanisms.

After a little algebra,

$$f_i(R_i) = \frac{\mu}{\lambda}$$

$$R_i = f_i^{-1}(\frac{\mu}{\lambda}) \qquad (4.14)$$

so that a monotonic function $f_i(R_i)$ generates a monotonic dependence of R_i on the shadow price ratio.

Setting $f_i(R_i) = \beta_i - \alpha_i R_i$—an 'exponential' model—then

$$R_i = \frac{\beta_i - \mu/\lambda}{\alpha_i} \qquad (4.15)$$

Sufficient environmental shadow price burden—selection pressure—will drive any individual modular regulatory channel capacity R_i below what is needed for critical function.

Shadow prices characterize selection pressures leading to decline in the maximum possible system control rate bandwidth R_{max} below some critical value R_{crit}. Optimization becomes impossible, and it should be possible to write R_{max} and R_{crit} for a critical subsystem as appropriate functions of some integral of the environmental shadow price over the duration of acute selection pressure. Details will vary across organisms and environments, with optimization model and measures of λ, μ and R, but the mechanism nevertheless seems canonical.

That is to say, shadow price burden will impose selection pressure in a similar manner for any possible scalar measure of fitness.

4.6 Discussion

Pivoting on the elegant work of Casanova and Konkel (2020), we have adapted models based on the asymptotic limit theorems of information and control theories to study the role of punctuated regulation in evolutionary punctuated equilibrium, generalizing matters far beyond genetic heritage mechanisms. The resulting treatment further expands, but remains consonant with, the Extended Evolutionary Synthesis, modulo an understanding of the many roles nonergodic information sources play at and across the varied scales and levels organization of the living state.

More generally, as in previous chapters, we have continued to trade the narrow-but-deep analysis of gene frequency models in evolutionary theory for broad-but-thin characterizations based on the asymptotic limit theorems of control and information theories. We have found in particular that 'punctuated equilibrium' dynamics emerge 'naturally' as a consequence of 'Fisher Zero' phase change and groupoid symmetry-breaking analogous to, but markedly different from, the more regular symmetry-breakings and related dynamics familiar from the physical theories of analogous phenomena.

Conversion of the various probability models explored here into reliable statistical tools for the analysis of observational and experimental data, however, remains to be done, and will be typically difficult, as will their reliable application.

4.7 Mathematical Appendix

We can examine the dynamics leading to Figs. 4.1 and 4.2 from an expanded perspective, taking the 'entropy' as

$$S \equiv -F(R) + R\,dF/dR - \frac{1}{2}R^2 d^2 F/dR^2 \tag{4.16}$$

and the gradient relation as

$$\partial R/\partial t = dS/dR - \frac{1}{2}Rd^2 S/dR^2 = \frac{R^3}{4}d^4 F/dR^4 \tag{4.17}$$

Then, for an 'exponential' model, the basic relations become

$$\frac{R^3}{4}d^4 F/dR^4 = \beta - \alpha R$$

$$F(R) =$$

$$\frac{C_2 R^2}{2} - 3 R^2 \alpha + 2 \ln(R) R\beta - 2 R\beta + \frac{C_1 R^3}{6} + 2\alpha R^2 \ln(R)$$

$$+ C_3 R + C_4$$

$$g = \frac{-F}{W(n, -F)} \tag{4.18}$$

where, again, $W(n, -F)$ is the Lambert W-function of order n.

Fig. 4.5 Analog to Fig. 4.1, setting $n = 0$, $\alpha = 1$, $C_1 = 1$, $C_2 = C_3 = -1$, $C_4 = 1/2$

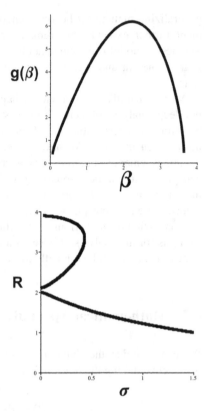

$g(\beta)$

β

Fig. 4.6 Analog to Fig. 4.2 for Fig. 4.5: the solution set $\{R, \sigma\}$ for $< dg_t >= 0$ in Fig. 4.3, taking $\beta = 2$. This contrasts greatly with Fig. 4.2, suggesting a means of differentiating underlying mechanisms

R

σ

These expressions can be explicitly solved in terms of α, β and four boundary conditions.

Figure 4.5 mirrors the signal transduction of Fig. 4.1, setting $n = 0$, $\alpha = 1$, $C_1 = 1$, $C_2 = C_3 = -1$, $C_4 = 1/2$.

Taking $\beta = 2$, the peak of Fig. 4.5, we can numerically calculate the solution set in $\{R, \sigma\}$ for $< dg_t >= 0$ via the Ito Chain Rule on the base relation of Eq. (4.7), producing Fig. 4.6, notably contrasting with Fig. 4.2. This may indicate a means of differentiating underlying mechanisms.

References

Britten, R., and E. Davidson, 1971. Repetitive and non-repetitive DNA sequences and a speculation on the origins of evolutionary novelty. Quarterly Review of Biology, 46:111–138.

Casanova, E., and M. Konkel, 2020. The developmental gene hypothesis for punctuated equilibrium: combined roles of developmental regulatory genes and transposable elements. Bioessays 42:e1900173.

Davidson, E., 2006. *The Regulatory Genome: Gene Regulation Networks in Development and Evolution*. Amsterdam: Elsevier.

de Groot, S., and P. Mazur, 1984. *Nonequilibrium Thermodynamics*. New York: Dover.

Dolan, B., W. Janke, D. Johnston, and M. Stathakopoulos, 2001. Thin Fisher zeros. Journal of Physics A 34:6211–6223.

Fisher, M., 1965. *Lectures in Theoretical Physics*, Vol. 7. Boulder: University of Colorado Press.

Gould, S.J., 2002. *The Structure of Evolutionary Theory*. Cambridge: Harvard University Press.

Horsthemeke, W., and R. Lefever, 2006. Noise-induced transitions. *Theory and Applications in Physics, Chemistry, and Biology*, Vol. 15. New York: Springer.

Jin, H., Z. Hu, and X. Zhou, 2008. A convex stochastic optimization problem arising from portfolio selection. Mathematical Finance 18:171–183.

Nair, G., F. Fagnani, S. Zampieri, and R. Evans, 2007. Feedback control under data rate constraints: an overview. Proceedings of the IEEE 95, 108138.

Nocedal, J., and S. Wright, 2006. *Numerical Optimization*, 2nd ed. New York: Springer.

Protter, P., 2005. *Stochastic Integration and Dierential Equations*, 2nd ed. New York: Springer.

Robinson, S., 1993. Shadow prices for measures of effectiveness II: general model. Operations Research 41:536–548.

Ruelle, D., 1964. Cluster property of the correlation functions of classical gases. Reviews of Modern Physics April:580–584.

Tian, C., L. Lin, and L, Zhang, 2017. Additive noise driven phase transitions in a predator-prey system. Applied Mathematical Modelling 46:423–432.

Van den Broeck, C., J. Parrondo, and R. Toral, 1994. Noise-induced nonequilibrium phase transition. Physical Review Letters 73:3395–3398.

Van den Broeck, C., J. Parrondo, J., R. Toral, and R. Kawai, 1997. Nonequilibrium phase transitions induced by multiplicative noise. Physical Review E, 55:4084–4094.

Weinstein, A., 1996. Groupoids: unifying internal and external symmetry. Notices of the American Mathematical Association 43:744–752.

Chapter 5
Institutional Dynamics Under Selection Pressure and Uncertainty

5.1 Introduction

Institutions are more than just complicated cultural artifacts. As Theiner et al. (2010) explain,

> Group cognition is not simply the unstructured aggregation of individual cognition, but the outcome of a division of cognitive labor among cognitive agents.

That is, institutions are cognitive entities in the sense of Atlan and Cohen (1998): an institution receives 'sensory' information from an environment that includes itself, compares that information with a learned or 'inherited', internal picture of the world—'doctrine,' 'policy,' 'business plan,' and the like—and then chooses a subset of the full set of responses possible to it. Choice reduces uncertainty, and the reduction of uncertainty implies the existence of an information source 'dual' to the cognitive action (Cover and Thomas 2006; Khinchin 1957; Wallace 2005, 2012, 2016, 2017, 2020, 2022a,b).

The argument is direct, intuitive, and implies that institutional dynamics taking place on 'selection landscapes' of fog, friction, and adversarial intent, are constrained by the asymptotic limit theorems of information theory (Dretske 1994; Wallace 2020). Such constraints lead toward probability models that might be reconfigured as robust statistical tools for the understanding and limited control of institutional dynamics under rapidly-changing selection pressures. As always, the design, testing, validation, and competent application of such tools would nonetheless be arduous. There is, after all, still no free lunch.

Cognition, it can be argued, must necessarily be paired with regulation under conditions of uncertainty and resource constraint. In consequence, we focus on an information theory model analogous to control theory's Data Rate Theorem characterizing the minimum necessary control information rate required to stabilize an inherently unstable system (Nair et al. 2007). Some development finds that external stress, as an economic 'shadow price' affecting optimization in the sense of Jin et al. (2008) and Robinson (1993), can force demand for rates of 'free energy' information and 'materiel' resources beyond feasible levels, driving damage accumulation and ultimate collapse. Related punctuated 'phase transitions' emerge in a remarkably natural, if somewhat subtle, manner.

Here, through worked-out examples, we project complicated and subtle control theory perspectives onto relatively simple information theory methods, using the standard approximate phenomenology of nonequilibrium thermodynamics (de Groot and Mazur 1984) and associated, if recondite, treatments of phase change involving 'groupoid symmetry-breaking' and 'Fisher zeros'.

5.2 A Rate Distortion Theorem Model of Control

Let H be the rate at which accurate information is available to an institution regarding the 'environment' in which it is embedded—including the effects of its own actions. Take C as the rate at which accurate information can be communicated between subcomponents within the institution, and M as the rate of supply of materiel resources, most often under conditions of loss and delay. Let Z be a resource rate scalarization constructed from the invariants of a 3×3 matrix representing direct and cross-interactions between C, H, and M. This might simply be taken as the product $Z = C \times H \times M$. The effect of Z is then opposed by a complicated 'noise' that represents environmental and other pressures.

A more detailed treatment of Z has been given in previous chapters, using a standard formalism to account for interactions between energy and information flows in a scalarized parameter. The Chapter Appendix also outlines the formalism necessary when Z must be taken as a more complicated algebraic object.

The basic control system feedback loop is shown in Fig. 5.1, a simplification of the usual 'OODA loop' of military theory (Osinga 2007; Wallace 2020).

A sequence of intent signals is transmitted from the left of the figure into the 'selection environment' represented by the rectangle. The synergistic measure of resource rate, Z, acts against a complicated background of multifactorial 'noise' that must be further parsed to become meaningful. This often does not happen. That is, a de-facto 'noise'—including adversarial actions and intent—degrades H, C, and M and their interactions, as represented in the scalarized rate measure Z. We assume a fixed time frame of observation, e.g., tactical, operational, or strategic, but not more than one.

Real-world effect—output—is then compared with intent –input—via the 'sensory' or 'intelligence' mechanisms of the dotted line, and a distortion measure D *on the time scale of interest* is constructed as a scalar index of disjunction, for example some composite of expected vs. observed production, failure rate, profit margin, total profit, employee turnover, stockholder satisfaction, and so on.

Let a sequence of a system's expectations-of-intent $X^i = x_1^i, x_2^i, \ldots$ be sent into the rectangle of Fig. 5.1 from the left, and compare it to the sequence of observed

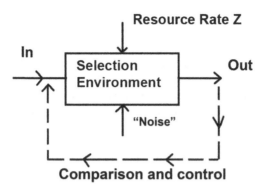

Fig. 5.1 A sequence of intent signals from an institution's command structure is transmitted on the timescale of interest via tactical, operational, and strategic actions into a 'selection environment' represented by the rectangle. The—possibly complicated—scalarized measure of resource delivery rate, $Z = Z(H, C, M)$, is seen as opposed by a 'noise' compounded of uncertainties, instabilities, imprecision-in-action, and active opposition. The actual effect is compared with the intent via mechanisms of the dotted line, and a 'distortion measure' D is taken as a scalar index of failure: the eye measures the gap as the hand grasps

outputs $\hat{X}^i$, i.e., what actually happens, to the right of the rectangle, via a distortion measure. Let $p(X^i)$ be the (here, assumed) probability of the sequence X^i.

The hand reaches, the eye measures the gap... and so on.

Then an average distortion D can be calculated as

$$D \equiv \sum_i p(X^i)d(X^i, \hat{X}^i) \tag{5.1}$$

where $d(X^i, \hat{X}^i)$ is a particular scalar measure of the distortion between the command input and the observed/measured result. Again, this might be a composite of expected vs. observed effect, gains, losses, precision-of-action, and so on. D might be sensed from repeated events as an ergodic measure in which time sequences are averaged to represent cross-sectional probability estimates. Here, we assume an estimate based on probabilities.

Note that defining the scalar distortion measure $d(X^i, \hat{X}^i)$ from multifactorial indices may require 'projection' methods similar to those used to derive the scalar resource rate Z.

It is possible *to interpret the process of turning X into $\hat{X}$ as the transmission of a signal along a noisy channel*, and we can impose a corresponding Rate Distortion Theorem argument (Cover and Thomas 2006). That theorem states that, under the conditions of Fig. 5.1, there is a minimum channel capacity $R(D)$ necessary to keep the average distortion at or below the scalar measure D.

Next, we construct a Boltzmann-analog pseudoprobability in the resource rate Z, using the argument by Feynman (2000) (and others) that information is itself a form of free energy. Feynman, basing an analysis on work by Bennett, constructs a clever ideal engine that can turn information into work. This approach leads to a standard expression for a probability density as

$$dP(R, Z) = \frac{\exp[-R/g(Z)]dR}{\int_0^\infty \exp[-R/g(Z)]dR} = \frac{\exp[-R/g(Z)]dR}{g(Z)} \tag{5.2}$$

The 'Rate Distortion temperature' $g(Z)$ must, however, be determined through the dynamic conditions imposed by the necessity of acting decisively under a time limit.

The analysis, to this point, is quite general. In what follows, however, we impose a 'worst case' example, the Gaussian channel (Shomorony and Avestimehr 2012), afflicted by Brownian white noise having a uniform spectral density. Most real-world noise will have 'color', that is, a unique spectral density, leading to relatively deep mathematical waters for general treatment (Protter 2005).

For such a Gaussian channel, under the squared distortion measure,

$$R(D) = (1/2) \log_2[\sigma^2/D]$$

where σ is a scalar noise measure. By convention, $R(D)$ is zero for $D \geq \sigma^2$.

Then $D = \sigma^2/4^R$ and some calculation defines a 'mean average' distortion $< D >$ as

$$< D >= \int_0^\infty \frac{\sigma^2}{4^R} dP(R, Z) = \frac{\sigma^2}{\log(4)g(Z) + 1} \tag{5.3}$$

The greater the 'temperature' $g(Z)$, the lower the distortion between intent and execution, and this is much the point.

Explicit determination of $g(Z)$ requires a dynamic model, and this is not trivial. Again, we follow Feynman (2000), and generalize the partition function of the denominator of Eq. (5.2), creating an *iterated free energy* F:

$$\exp[-F/g(Z)] = \int_0^\infty \exp[-R/g(Z)]dR = g(Z)$$

$$F = -g(Z) \log[g(Z)]$$

$$g(Z) = \frac{-F}{W(n, -F)} \tag{5.4}$$

where $W(n, x)$ is the Lambert W-function of order n solving the relation

$$W(n, x) \exp[W(n, x)] = x$$

$W(n, x)$ is real-valued only for $n = 0, -1$ over the respective ranges $x \geq -\exp[-1]$ and $-\exp[-1] \leq x \leq 0$.

The development allows, in turn, definition of an entropy-analog as the Legendre transform (Pettini 2007) of F,

$$S \equiv -F(g(Z)) + Z dF(g(Z))/dZ \tag{5.5}$$

We then use that 'entropy' in an extension of the Onsager approximation of nonequilibrium thermodynamics (de Groot and Mazur 1984), taking the time rate of change of Z as (approximately, in first-order) proportional to the gradient of S in Z:

$$dZ/dt \approx \mu dS/dZ = f(Z)$$
$$Z d^2 F/dZ^2 = f(Z) \tag{5.6}$$

We will incorporate the 'diffusion coefficient' μ into the function $f(Z)$ and assume that $f(Z) \to 0$ as $t \to \infty$.

$f(Z)$ might be, for example, the 'exponential' expression $dZ/dt = \beta - \alpha Z(t)$, so that $Z(t) \to \beta/\alpha$, the 'Arrhenius' model, $Z(t) = \beta \exp[-\alpha/t]$, with $f(Z) = (Z/\alpha)(\log(Z/\beta))^2$, and so on.

The differential equations that emerge from Eqs.(5.4–5.6) can easily be solved to give

$$X \equiv -C_1 Z - Z \int \frac{f(Z)}{Z} dZ + C_2 + \int f(Z) dZ$$
$$g(Z) = \frac{X}{W(n, X)} \tag{5.7}$$

Again, the $W(n, X)$ are Lambert W-functions of orders $n = 0, -1$.

$g(Z)$ must be real-valued positive and monotonic increasing in Z for the distortion relation of Eq. (5.3) to work. Relaxing that requirement leads to bifurcations at critical values of Z, when $g(Z)$ shifts from being real for $n = 0$ to being real for $n = -1$. Again, for simplicity, the diffusion coefficient μ has been absorbed into $f(Z)$.

The lack of microreversibility in information processes implies that multidimensional versions of the theory do not enjoy 'Onsager reciprocal relations', due to the inherent one-way 'directed homotopy' associated with information transmission. Such dihomotopy will be found to have important implications for the symmetries underlying punctuated cognitive dynamics, as opposed to 'simple' physical phase transition.

In particular, there may be circumstances and values of Z for which a positive, monotonic increasing $g(Z)$ ceases to be real—essentially analogs to the 'Fisher zero' phase transitions from physical theory (Dolan et al. 2001; Fisher 1965; Ruelle 1964). Directed homotopy provides a simple bifurcation model of groupoid symmetry-breaking phase transitions (Golubitsky and Stewart 2006), and this is important. See the Mathematical Appendix of Chap. 1 for an introduction to groupoid theory, representing the generalization of group symmetry-breaking arguments from physical theory into cognitive theory.

We will return to these matters below.

The observables are the distortion measure $< D >$ and the resource rate Z. The free energy analog F and the derived entropy analog S are intermediate variables that, in this model, allow characterization of the symmetry-breaking dynamics of $< D >$ as driven by Z through the properties of $g(Z)$ and the real-value limitations of the Lambert W-function.

This approach is roughly consonant with analogous results from physical theory, much as quantum mechanics is based on replacing variables in a classical mechanics Hamiltonian with operators characterized by highly punctuated spectra, often dominated by group symmetry properties. The 'spectra' here are the groupoid symmetry-breaking punctuations in the 'temperature' $g(Z)$. Again, this is consistent with 'Fisher zero' treatments of phase transition involving the appearance of complex-valued temperatures.

5.3 Selection Pressure Dynamics

A First Model

It is possible to model the influence of selection pressure on a system dominated by the distortion-temperature relation of Eq. (5.3) under conditions of both resource and time limitations. Such constraint most particularly involves a limit to the rate of supply of the materiel and information resources needed to maintain function under selection pressure.

The full system is seen as consisting of $i = 1, 2, ..., m$ cooperating units that must be individually supplied with resources at rates Z_i under the overall constraints of resources and time as $\sum_i Z_i = Z$, $\sum_i T_i = T$. We seek to minimize the total distortion between intent and effect according to a Lagrangian constrained optimization taking $dZ_i/dt = f(Z_i) \to 0$. This is, perhaps, the simplest entry into the dynamics of optimization. Other approaches abound (e.g., Nocedal and Wright 2006).

Although, from Eq. (5.3), we could take $D_i = \sigma^2/[\log(4)g(Z_i) + 1]$, we can construct—or impose—a more general Lagrangian optimization across all possible definitions of distortion as

$$\mathscr{L} = \sum_i D_i + \lambda \left(Z - \sum_i Z_i \right) + \mu \left(T - \sum_i T_i \right)$$

$$\partial \mathscr{L}/\partial Z = \lambda$$

$$\partial D_i/\partial Z_i = \lambda$$

$$\partial D_i/\partial T_i = \mu \qquad (5.8)$$

remembering in particular that $\partial Z_i/\partial T_i = f_i(Z_i)$.

In simple economic models, λ and μ represent the 'shadow prices' imposed on system dynamics by environmental constraints, in a large sense (Jin et al. 2008; Robinson 1993).

Again, other formalisms are possible (e.g., Nocedal and Wright 2006), but a simple Lagrangian optimization outlines what is to be expected from any realistic optimization strategy. The assertion is that other optimization methods will, like other forms of rate distortion functions, display an underlying canonical similarity (Cover and Thomas 2006).

Some 'elementary' manipulation gives

$$f_i(Z_i) = \frac{\mu}{\lambda}$$

$$Z_i = f_i^{-1}(\frac{\mu}{\lambda}) \tag{5.9}$$

so that a monotonic function $f_i(Z_i)$ generates a monotonic dependence of Z_i on the shadow price ratio.

For example, setting $f_i(Z_i) = \beta_i - \alpha_i Z_i$—an 'exponential' model—produces the expression

$$Z_i = \frac{\beta_i - \mu/\lambda}{\alpha_i} \tag{5.10}$$

Sufficient environmental shadow price burden will drive any resource rate Z_i below what is needed for critical modular function under selection pressure.

Again, λ and μ are understood from arguments in economics to represent the 'shadow prices' imposed by embedding environmental constraints. As Jin et al. (2008) show, if the shadow prices—environmental stress measures—are beyond what the system can meet, then optimization fails.

Shadow prices characterize environmental burdens that accelerate attrition, leading to decline in the maximum possible level of resource delivery Z_{max} below some critical value Z_{crit}. Optimization becomes impossible, and we could then write both Z_{max} and Z_{crit} for a critical subsystem as appropriate functions of some integral of the environmental shadow price over the duration of acute selection pressure. Details will vary across institutions and environments, with optimization model and measures of λ and Z, but the mechanism seems canonical, potentially applying across many systems.

A Second Model

It is possible to reformulate this analysis in terms of a rate of modular or large-scale institutional cognition, using Eq. (5.2) to calculate a classic rate-of-reaction model (Laidler 1987) in terms of the probability that $R \geq K$:

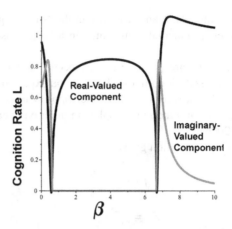

Fig. 5.2 Classic inverted-U signal transduction form for the rate of cognition vs. 'arousal' β. Here, based on Eq. (5.7), $n = 0$, $\alpha = 1$, $C_1 = C_2 = -3$, $K = 1$. Only for a limited range of β is the cognition rate simply real-valued. Beyond that range, nonzero imaginary-valued components introduce gross instability, consistent with 'Fisher zero' treatment of phase change in physical systems

$$L \equiv P(R \geq K) = \frac{\int_K^\infty \exp[-R/g(Z)]dR}{\int_0^\infty \exp[-R/g(Z)]dR} = \exp[-K/g(Z)] \qquad (5.11)$$

$g(Z)$ remains as in Eq. (5.7), and the optimization/shadow price arguments are 'retransmitted', in a sense, through this expression.

In Fig. 5.2 we assume an exponential model $dZ/dt = f(Z) = \beta - \alpha Z$ and construct a Yerkes-Dodson 'arousal'/'signal transduction' graph (Diamond et al. 2007) for the cognition rate L, using the $n = 0$ branch of the Lambert W-function. That is, we fix α. Here, $Z(t) = (\beta/\alpha)(1 - \exp[-\alpha t])$, which varies while β increases as the Yerkes-Dodson 'arousal' measure. Using Eq. (5.7), we take $n = 0$, $\alpha = 1$, $C_1 = C_2 = -3$, $K = 1$. Note the classic inverted-U 'dose-response' signal transduction pattern for cognition rate vs. arousal. Outside a limited range of arousal β, the system suffers non-zero imaginary-valued components, introducing gross temporal instabilities consistent with 'Fisher zero' phase transitions involving occurrence of complex-valued temperatures.

The deterministic relation of Eq. (5.6) can be expanded for analysis of the stability of institutional cognition under 'random noise' in the delivery of essential resources, as opposed to noise in the transmission of information. Equation (5.6) stated that $dZ/dt \approx f(Z)$ for some appropriate function. That is, it takes time for a cognitive system to respond to resource demands. There is always some delay, that will inevitably interact with other factors. As an example, we take the exponential model, i.e., $f(Z) = \beta - \alpha Z(t)$, $Z(t) \to \beta/\alpha$. α determines the rate at which changes in demand can be met.

We first explore the effect of 'noise' on the stability of Z via a stochastic differential equation (SDE) in standard form (Protter 2005):

$$dZ_t = (\beta - \alpha Z_t)dt + \sigma Z_t dB_t \tag{5.12}$$

where the second term represents volatility proportional to Z itself, an effect driven by the parameter σ. dB_t is here taken as ordinary Brownian white noise. Again, this is the usual starting point for such analyses, which can be extended to 'colored' noise having a nonuniform spectral density via the Doleans-Dade exponential (Protter 2005).

It is surprisingly easy to examine the second order stability of this system by invoking the Ito Chain Rule on Z^2 (Protter 2005), calculating the variance of Z as

$$< Z^2 > - < Z >^2 = \left(\frac{\beta}{\alpha - \sigma^2/2}\right)^2 - \left(\frac{\beta}{\alpha}\right)^2 \tag{5.13}$$

This expression explodes as $\sigma^2/2 \to \alpha$, a condition imposing itself regardless of the magnitude of $Z_\infty = \beta/\alpha$.

Integrated damage accumulation will affect both Z_∞ and σ as logistic and supervisory systems either cannot compensate or progressively fail.

Next, we extend the $g(Z)$ models to include stochastic stability effects, applying the Ito Chain Rule (Protter 2005) to the cognitive signal transduction model of Fig. 5.2 in the presence of other sources of accumulated stress, in addition to β 'arousal' effects. The development is surprisingly straightforward.

A Third Model

Again taking the underlying SDE for Z as Eq. (5.12), we apply the Ito Chain Rule across the cognition rate L from Eq. (5.11), calculating the nonequilibrium steady state as $< dL_t > = 0$. Figure 5.3 shows a typical result for the solution set $\{Z, \sigma\}$. Here, increasing 'noise' σ rapidly degrades the de-facto available resource rate Z.

We can take the argument one step further, imposing an index of chronic selection stress Ω according to the more general SDE

$$dZ_t = (f(Z_t) + \Omega(Z_t))dt + \sigma h(Z_t)dB_t \tag{5.14}$$

The relation for cognition rate, $< dL_t > = 0$, taking $L = \exp[-K/g(Z)]$, is then solved via the Ito Chain Rule in terms of $g(Z)$ as

Fig. 5.3 Solution to the nonequilbrium steady state condition for cognition rate, $< dL_t >= 0$ at the real-valued peak of Fig. 5.2. Here, $\alpha = 1$, $\beta = 4$, $C_1 = C_2 = -3$, $K = 1$. Increasing 'noise' σ rapidly degrades the effective level of Z

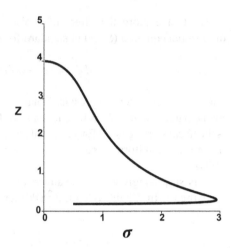

$$g(Z) = K\left(\ln\left(\frac{1}{K}\left(C_1\int\left(e^{\frac{1}{\sigma^2}\int\frac{f(Z)}{(h(Z))^2}+\frac{\Omega(Z)}{(h(Z))^2}dZ}\right)^{-2}dZ + C_2\right)^{-1}\right)\right)^{-1} \quad (5.15)$$

Crudely, if $\Omega(Z_t)$ is positive—for example 'Fuhrer Befiehl' micromanagement, generalized supervisory incompetence, widespread employee dissatisfaction, hypercompetent competition, and the like—then Fig. 5.2 is driven to the right. If $\Omega(Z_t)$ is negative—employee burnout/turnover, supply chain/delivery/production problems—the system is driven to the left.

Chronic selection stress and/or attrition can become synergistic with uncertainty in causing gross instability.

5.4 Destabilization By Delay

A significant complication arises from overtly delayed response to stress, in addition to generalized uncertainties and other burdens.

Suppose some fixed delay $\delta \geq 0$ is imposed, so that the deterministic and stochastic dynamic relations become

$$dZ/dt = f(Z(t-\delta))$$
$$dZ_t = f(Z_{t-\delta})dt + \sigma h(Z_t)dB_t \quad (5.16)$$

Assuming an exponential model $f(Z) = \beta - \alpha Z(t-\delta)$, relatively straightforward algebra gives

$$Z(t) = \frac{\beta}{\alpha}(1 - \exp[St])$$

$$S = S(n, \alpha\delta) \equiv \frac{W(n, -\alpha\delta)}{\delta}$$

$$dZ/dt = S(n, \alpha\delta)(Z(t) - \beta/\alpha)$$

$$dZ_t = S(n, \alpha\delta)(Z_t - \beta/\alpha)\,dt + \sigma Z_t dB_t \qquad (5.17)$$

Again, dB_t is Brownian noise and $W(n, x)$ is the Lambert W-function of order n. Recall that W is real-valued only for orders $n = 0, -1$ and over the limited ranges $x \geq -\exp[-1]$ and $-\exp[-1] \leq x \leq 0$. Consequently, if $\delta\alpha > \exp[-1]$, the system will have nonzero imaginary components: the dynamic behavior becomes unstable, often producing quasi-sinusoidal oscillation.

Analogous to, but different from, Eq. (5.13), for the last expression of Eq. (5.17), the SDE in Z is stable in second order only if $\|S\| > \sigma^2/2$, recalling that S must be real-valued and negative for stability. That is, as above, sufficient 'noise' can destabilize an otherwise stable delayed system. These results will be generalized in the next section.

An interesting example. There is no a priori reason to assume one particular real-valued deterministic solution for Eq. (5.17) over another. Consequently, both modes may become independently manifest and we use the *average* of the $n = 0$ and $n = -1$ forms. Figure 5.4 shows that average, with $\alpha = 1$, $\beta = 2$, $\delta = 3/2$. If $\alpha\delta \gg \exp[-1]$, some consideration finds the $n = 0$ and $n = -1$ Lambert W-modes of $Z(t)$ differ only by the sign of the imaginary component, so that the average is real-valued and oscillatory, converging to the limit β/α only if $\alpha\delta < \pi/2$. This behavior is closely analogous to vehicle 'fishtailing' triggered by response delays and subsequent overcorrections.

5.5 Extending the Data Rate Theorem

The Data Rate Theorem (Nair et al. 2007) describes how stabilization of an inherently unstable system requires that control information be imposed at a rate greater than the sources of instability can generate their own 'topological information'. Envision a vehicle driven at a particular speed on a twisting, pot-holed roadway at night. Steering, transmission dynamics, braking, headlights, and driver skill must impose control information at a rate greater than the twists and bumps of the roadway can impose themselves on the vehicle at its chosen speed. One common vehicle failure mode is via the fishtailing mechanism described in the previous section.

General mathematical proofs of the DRT are usually based on an extension of the Bode Integral Theorem. Here, we are not concerned only with control information, but with the function of the institutional system *as determined by the scalar resource*

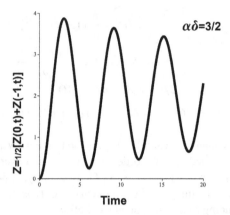

Fig. 5.4 Delayed overcorrection in system dynamics. From Eq. (5.17), since there is no particular preference for solution form, both modes may be active. Here, we take the average $Z(t) = 1/2[Z(0, t) + Z(-1, t)]$ across Lambert W-functions of orders $n = 0, -1$, the only ones having real-valued components. $\alpha = 1$, $\beta = 2$, $\delta = 3/2$ so that $\alpha\delta \gg \exp[-1]$. Fishtailing oscillation instability becomes manifest, eventually converging to β/α only if $\alpha\delta < \pi/2$. (Wallace 2022a, Ch. 9) uses similar methods to explain recurrent waves of COVID-19 infections following failure of institutional cognition and strategic planning in US and other public health systems

rate index Z. Via a different free energy argument than above, much will be made of the inherent convexity of all Rate Distortion Functions (Cover and Thomas 2006), that is, the fact that $dR^2(D)/dD^2 \geq 0$.

Again following Feynman (2000), we now characterize the channel capacity defined by the Rate Distortion Function $R(D)$ as itself a 'free energy' measure, permitting definition of another 'entropy' via a Legendre transform in the distortion D,

$$S \equiv -R(D) + DdR/dD \qquad (5.18)$$

The deterministic first order Onsager nonequilibrium thermodynamics relation is then

$$dD/dt \propto dS/dD = Dd^2R/dD^2 \qquad (5.19)$$

recognizing that, by convexity $d^2R/dD^2 \geq 0$.

For the systems of interest in this work, the appropriate stochastic differential equation has the form

$$dD_t = (D_t d^2R/dD_t - M(D))dt + \Omega h(D_t)dB_t \qquad (5.20)$$

Here, $M(D) \geq 0$ is the 'countervailing force', the 'control signal', needed to stabilize the system, and the second term is a measure of 'volatility' in the white noise dB_t.

We are interested in 'Q-order stability' defined by some function $Q(D)$. Most simply, this might be second order stability in variance, so that $Q(D) = D^2$, but the argument is more general. Q-order stability can be found using the Ito Chain Rule to calculate the nonequilibrium steady state $< dQ_t >= 0$. The base equation is then

$$(Dd^2 R/dD^2 - M(D))dQ/dD + \frac{\Omega^2}{2}h(D)^2 d^2 Q/dD^2 = 0 \tag{5.21}$$

Solving for $M(D)$

$$M(D) = \frac{\Omega^2 (h(D))^2 \frac{d^2}{dD^2} Q(D)}{2 \frac{d}{dD} Q(D)} + D\frac{d^2}{dD^2} R(D)$$

$$M(D) > \frac{\Omega^2 (h(D))^2 \frac{d^2}{dD^2} Q(D)}{2 \frac{d}{dD} Q(D)} \tag{5.22}$$

remembering that, by convexity, $d^2 R/dD^2 \geq 0$.

This result generalizes the Data Rate Theorem beyond the simple 'control information' rate.

Assuming $h(D) = D$—classic volatility—and setting $Q(D) = D^2$ gives the condition for second order stability for a Gaussian channel, having $R(D) = (1/2) \log_2(\sigma^2/D)$, as

$$M(D) = \frac{\Omega^2}{2} D + \frac{1}{D \log(4)}$$

$$M(D) \geq \frac{\Omega}{\sqrt{\log(2)}} \tag{5.23}$$

where the second expression is given in terms of the absolute minimum of the first.

5.6 Moving On

The simple stochastic exponential model,

$$dZ_t = (\beta - \alpha Z_t)dt + \sigma Z_t dB_t \tag{5.24}$$

where dB_t is Brownian noise, approaches a nonequilibrium steady state provided Z at time zero $\neq 0$ and the second order stability condition $\alpha > \sigma^2/2$ is satisfied.

If, however, the underlying relation can be written as $dZ_t = Z_t dY_t$, where dY_t is itself a stochastic differential equation—not necessarily driven by Brownian white noise—then a general solution, called the Doleans-Dade exponential (Protter 2005) can be found as

$$Z_t \propto \exp(Y_t - 1/2[Y_t, Y_t]) \tag{5.25}$$

where $[Y_t, Y_t]$ is the quadratic variation of Y_t (Protter 2005). If $[Y_t, Y_t] > 2Y_t$, then Z converges in probability to zero.

There is a similar general result, following Appleby et al. (2008). Suppose Z is of dimension ≥ 2, so that the base deterministic relation is $d\mathbf{Z}/dt = f(\mathbf{Z})$, with f now a vector function. Appleby et al. (2008) show that, under appropriately general circumstances, for the SDE

$$d\mathbf{Z}_t = f(\mathbf{Z}_t)dt + h(\mathbf{Z}_t)d\mathbf{B}_t \tag{5.26}$$

where h and $\mathbf{B}_t$ are also of dimension $n \geq 2$, a 'noise function' $h(\mathbf{Z}_t)$ can always be found such that a nonzero nss stable configuration—one having $f(\mathbf{Z}_0) = 0$, $\mathbf{Z}_0 \neq 0$—is nonetheless driven to zero in probability. That is, $\mathbf{Z}_t \to 0$.

In sum, consonant with the 'shadow price' optimization results above, for complex multidimensional systems, there will always be some selection stress manifested as a 'noise' load that is sufficient to collapse the system.

5.7   Reconsidering Cognition *An Sich*

The centrality of the previous analysis is the channel connecting institutional intent with institutional execution, characterized by a Rate Distortion Function $R(D)$, as expressed by the Morse Function defined through Eq. (5.4). This approach requires introducing the temperature analog of Eq. (5.7). An institutional cognition rate was then defined through Eq. (5.11), again in terms of the channel linking intent with effect, leading to the complicated phase transitions represented in Fig. 5.2.

By some contrast, this section focuses on cognition and cognitive dynamics as seen from within the institution, placing effect, as it were, in the hands of God. This is a markedly different perspective, leading to a different view of failure of cognition under stress. That is, for this development, the internals of the institution are most explicitly manifest.

The basic idea centers on the recognition by Atlan and Cohen (1998) that the essence of cognition is the choice of action or actions from a larger set of those available. Choice reduces uncertainty, and the reduction of uncertainty implies the existence of an information source dual to the cognitive process under study. Cognitive phenomena on dynamic landscapes of fog, friction, and adversarial intent, will not usually be ergodic—cross sectional means corresponding to temporal averages. Following Khinchin (1957), while an information source uncertainty may defined for an individual behavioral path, the value of the uncertainty varies between paths and cannot be defined as a 'Shannon entropy' across a fixed probability set. The reduction of a nonergodic source to a component sum of ergodic processes, in our context, is like parsing Keplerian orbital mechanics into Fourier sums of perfect circle Ptolemaic epicycles. The basic thing has been lost.

Institutions are composed of cognitive submodules interacting by crosstalk, as constrained by the embedding culture and environment, including the actions and intents of competing and cooperating institutions.

Further, there is always a structured uncertainty imposed by the large deviations most probable within that environment.

Thus, a number of factors interact to build a composite information source (Cover and Thomas 2006) representing institutional cognition:

1. As stated, cognition requires choice reducing uncertainty, implying an information source X_i 'dual' to cognition at each scale and level of organization (Atlan and Cohen 1998). The argument is direct, compelling, and intuitive.
2. Embedding culture also imposes an information source Ω, with analogs to grammar and syntax. That is, within a culture, under particular circumstances, some sequences of behaviors are highly probable, and others have vanishingly small probability, a sufficient condition for the nonergodic formalism (Khinchin 1957).
3. Embedding spatial and social geographies, characterized by an information source V, similarly have incident sequences of very high and very low probability. For example, night follows day, summer's dirt roads are followed by October's impassible mud, and so on.
4. Large deviations (Champagnat et al. 2006; Dembo and Zeitouni 1998) follow high probability developmental pathways governed by entropy-like laws that imply the existence of another information source L_D.
5. Cognitive processes on dynamic 'roadways' are almost always paired with regulatory information sources X^i: immune systems must not attack self-tissues, consciousness must be paired with social control.

Institutional cognition, like other embedded and embodied forms of cognition, can thus be characterized by a joint information source uncertainty (Cover and Thomas 2006):

$$H(\{X_i, X^i\}, \Omega, V, L_D) \tag{5.27}$$

Again, this information source is not likely to be ergodic. To reiterate, nonergodic information sources cannot be characterized as 'entropies', having the form $-\sum_j P_j \log[P_j]$ where the P_j are across some probability distribution. See Khinchin (1957) for a detailed discussion. For nonergodic information sources, however, one can assign a *path dependent* information source uncertainty.

The set $\{X_i, X^i\}$ is taken to pair the basic cognitive process X_i with a needed regulatory process X^i for essential subcomponent modules, i.e., the 'command structure', in a large sense, will be constrained by 'doctrine'. Again, for biological systems, major cognitive phenomena are almost always paired with essential regulators, as in gene expression and immune function across the life course. Failure of such bioregulators in higher animals is most often associated with the onset of the diseases of aging.

Conversely, however, as the Soviet Union, and now Russia, have found in a number of episodes, too-rigid enforcement of military doctrine—an army's corporate genome—can impede tactical flexibility and the ability to respond to other-scale institutional selection pressures. Institutional ecologists in business schools may speak of 'buggy-whip industries'. Wallace (2015) speaks to some of this.

We have, above, reduced the spectrum of resources and their interactions, including internal bandwidth, rates of sensory information, and material supply, to a scalar rate variable Z.

We can now construct another—more detailed—iterated free energy Morse Function (Pettini 2007) characterizing a particular institution, using a Boltzmann probability expression—again in the sense of Feynman (2000). The first step is to enumerate the high probability developmental pathways available to the system. Taking $j = 1, 2, ...$, it is possible to define probability P_j for a particular path j as

$$P_j = \frac{\exp[-H_j/g(Z)]}{\sum_k \exp[-H_k/g(Z)]} \tag{5.28}$$

The analysis applies to nonergodic as well as to ergodic information sources and can be used for systems in which each developmental pathway x_j has its own source uncertainty measure H_{x_j}. Again, this value can only be defined as a 'Shannon entropy' for an ergodic system.

The 'temperature' $g(Z)$ will be calculated from Onsager-like system dynamics built from the partition function, that is, from the denominator of Eq. (5.28).

An institutional cognition rate can then be expressed via a standard chemical reaction theory (Laidler 1987). The rate is given by the probability such that $H_j > H_0$, where H_0 is the lower limit for detection of a signal in a varying and noisy environment.

Another Morse Function F can be written as

$$\exp[-F/g(Z)] \equiv \sum_k \exp[-H_k/g(Z)] \equiv h(g(Z))$$

$$F = -\log[h(g(Z))]g(Z)$$

$$g(Z) = \frac{-F}{RootOf\,(\exp[Y] - h(-F/Y))}$$

$$\tag{5.29}$$

If $h(g(Z)) = g(Z)$ then $g(Z) = -F/W(n, -F)$, where $W(n, x)$ is the Lambert W-function of order n that satisfies the relation $W(n, x)\exp[W(n, x)] = x$. It is real-valued only for $n = 0, -1$ respectively over the ranges $x \geq -\exp[-1]$ and $-\exp[-1] \leq x \leq 0$.

The Root of construction should thus be interpreted as a generalized Lambert W-function.

There are several matters of interest.

1. The 'RootOf' construction may have complex number solutions so that the 'temperature' function $g(Z)$ is analogous to the more common 'Fisher Zeros' that characterize phase transitions in physical systems (Dolan et al. 2001; Fisher 1965; Ruelle 1964).
2. Since information sources are fundamentally dissipative—palindromes are vanishingly rare and directed homotopy dominates—microreversibility is impossible. As a direct consequence, there can be no 'Onsager Reciprocal Relations' in higher dimension systems.
3. F is a Morse Function that is subject to symmetry-breaking transitions as $g(Z)$ varies (Pettini 2007). The symmetries here, however, are not those of 'simple' physical phase transitions, most often represented by standard group structures. Cognitive phase change involves punctuated transitions between equivalence classes of high probability directed signal sequences, necessarily represented as groupoids. These are extensions of groups in which a product is not necessarily defined for every possible pair of elements (Brown 1992; Cayron 2006; Weinstein 1996). The emergence of groupoids appears to be a consequence of the inherently one-way directed homotopy of information sources.

To reiterate in the context of cognition, groupoid symmetries are driven by the directed homotopy induced by failure of local time reversibility for information systems. Again, this occurs because palindromes have vanishingly small probability: in English, ' the ' has meaning in context while ' eht ' has vanishingly low probability.

More complicated cognitive systems may even require more general structures, for example, small categories or even semigroupoids, for analogs to the standard symmetry-breaking dynamics of physical systems.

There may, then, be a number of phase analogs available to a cognitive system as $g(Z)$ varies, rather than just the 'on/off' of stability implied by the Data Rate Theorem.

Again, dynamic behavior of institutional cognition can be derived via another Onsager approximation in the gradient of an iterated entropy-like measure constructed from the iterated free energy Morse Function F via a Legendre transform, in a now familiar manner (de Groot and Mazur 1984). Recall the results

$$\exp[-F/g(Z)] =$$

$$\sum_k \exp[-H_k/g(Z)] \equiv h(g(Z))$$

$$F(Z) = -\log(h(g(Z))g(Z)$$

$$S(Z) \equiv -F(Z) + Z dF(Z)/dZ$$

$$\partial Z/\partial t \approx \mu \partial S/\partial Z = f(Z)$$

$$f(Z) = Z d^2 F/dZ^2 \tag{5.30}$$

absorbing μ into $f(Z)$.

$f(Z)$ determines the rate at which the system can adjust to changes in Z.

The RootOf calculation gives, in a first-order Onsager-analog approximation, the implicit equation

$$-Z \int \frac{f(Z)}{Z} dZ - \log(h(g(Z)))g(Z) -$$

$$C_1 Z + \int (f(Z)dZ + C_2 = 0 \tag{5.31}$$

leading back to the last expression of Eq. (5.29).

Again, $F = -\log(h(g(Z))g(Z)$, but h is, in general, hard to make explicit, since it is determined by the details of institutional internal structures. We will give a 'simple' example below. In one sense, this result most differentiates the 'cognitive' from the previous 'distortion' approach.

Note that specifying of any two of f, g, h, allows calculation of the third. Again, h is determined by the internal structure of the larger system, f is imposed by 'frictional' externalities, and the boundary conditions C_1, C_2 are likewise imposed by externalities. Thus the temperature-analog $g(Z)$ is environmentally determined, in a large sense.

We have assumes that the cognitive institution under study can be characterized by a single scalar parameter Z that mixes material resource/energy supply with internal and external flows of information under time constraint. There may be more than one such irreducible composite driving system dynamics. That is, it may be necessary to replace the scalar Z with an $m \leq n$-dimensional vector having a number of independent—even orthogonal—components accounting for significant portions of the total variance in the rate of supply of essential resources. The dynamic equations must then be presented in more complicated vector form, as described in previous chapters.

A 'simple' example.

We suppose a two-level system, above and below a 'detection energy' H_0 by amounts δ, so that the 'partition function' characterization of 'free energy' and cognition rate become

$$\exp[-F/g(Z)] = \exp[-H_0/g(Z)] \left(\exp[-\delta/g(Z)] + \exp[\delta/g(Z)] \right) =$$

$$\exp[-H_0/g(Z)] 2 \cosh[\delta/g(Z)]$$

$$F(Z) = -\log[2\cosh(\delta/g(Z))]g(Z) + H_0$$

$$L \equiv Pr[H > H_0] = \frac{\exp[-(H_0 + \delta)/g(Z)]}{\exp[-(H_0 + \delta)/g(Z)] + \exp[-(H_0 - \delta)/g(Z)]} =$$

$$\frac{1}{1 + \exp[2\delta/g(Z)]} \tag{5.32}$$

As above, the first order Onsager approximation is

$$S = -F(Z) + Z dF/dZ$$

$$dZ/dt = f(Z) = dS/dZ = Z d^2 F/dZ^2 \tag{5.33}$$

Reexpressing F as $F(g(Z))$ gives a differential equation for $g(Z)$ as

$$\frac{\delta \left(\frac{d^2}{dZ^2} g(Z)\right) \sinh\left(\frac{\delta}{g(Z)}\right)}{g(Z) \cosh\left(\frac{\delta}{g(Z)}\right)} - \frac{\delta^2 \left(\frac{d}{dZ} g(Z)\right)^2}{g(Z)^3} +$$

$$\frac{\delta^2 \left(\frac{d}{dZ} g(Z)\right)^2 \left(\sinh^2\left(\frac{\delta}{g(Z)}\right)\right)}{g(Z)^3 \cosh\left(\frac{\delta}{g(Z)}\right)^2} -$$

$$\ln\left(2 \cosh\left(\frac{\delta}{g(Z)}\right)\right) \left(\frac{d^2}{dZ^2} g(Z)\right)$$

$$= f(Z)/Z \tag{5.34}$$

Approximating Eq. (5.34) to third order in δ,

$$-\ln(2)\left(\frac{d^2}{dZ^2} g(Z)\right) + \left(\frac{\frac{d^2}{dZ^2} g(Z)}{2g(Z)^2} - \frac{\left(\frac{d}{dZ} g(Z)\right)^2}{g(Z)^3}\right) \delta^2 \approx f(Z)/Z \tag{5.35}$$

There are two solutions, differing by sign in a second term. Since we cannot differentiate between them a priori, we can take the full solution as their average. First in general, and then imposing an exponential model as $f(Z) = \beta - \alpha Z$,

$$g(Z) \approx \frac{-2Z \int \frac{f(Z)}{Z} dZ - C_1 Z + 2 \int f(Z) dZ + C_2}{4 \ln(2)}$$

$$g(Z) \approx \frac{-2 \ln(Z) Z\beta + \alpha Z^2 + (-C_1 + 2\beta) Z + C_2}{4 \ln(2)} \tag{5.36}$$

Recalling the expression for the cognition rate L for the two-level system, and setting $\alpha = 1$, $C_1 = C_2 = 0$, $\delta = 1/3$, gives the 'arousal' graph of Fig. 5.5. Note the cognitive phase transition that occurs when $g(\beta) \leq 0$.

Again, the Yerkes-Dodson law appears, followed, at a critical value of arousal, by the onset of thrashing/panic, but, in this particular case, without emergence of an imaginary-valued component.

Figure 5.6 carries out the stochastic analysis of L in Eq. (5.32), via the Ito Chain Rule applied through Eq. (5.12), similar to Fig. 5.3. Here, $\alpha = 1$, $\beta = 5/2$, $C_1 = C_2 = 0$, $\delta = 1/3$, as in Fig. 5.5, taking β at the peak of the inverted U. Again, the

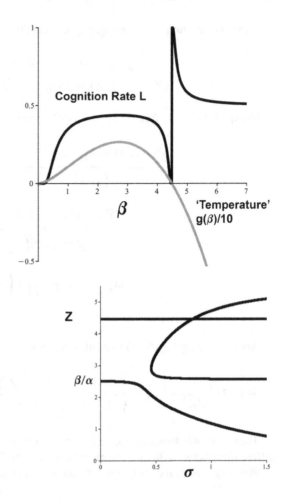

Fig. 5.5 Cognition rate L and 'temperature' $g(\beta)/10$ for the two-level cognitive system. This is an analog to Fig. 5.2. Here, $\alpha = 1$, $C_1 = C_2 = 0$, $\delta = 1/3$. Again, the inverted-U Yerkes-Dodson law emerges, followed by a sudden phase transition to 'panic'—when $g \le 0$—at a critical arousal level β

Fig. 5.6 The equivalence class set in $\{Z, \sigma\}$ for the relation $< dL_t >= 0$ from Fig. 5.5, taken at the peak of the inverted-U. Note the initial bifurcation at $\sigma \approx 0.457$, much less than the instability limit in Z itself at $\sigma = \sqrt{2}$. The upper straight line represents the stability limit, ≈ 4.45 from Fig. 5.5, reached at $\sigma \approx 0.82$. Other parameter values are those of Fig. 5.5, again fixing $\beta = 5/2$

figure is the equivalence class set in $\{Z, \sigma\}$ for the relation $< dL_t >= 0$. Note the bifurcation at $\sigma \approx 0.457$, much less than the instability limit in Z itself at $\sigma = \sqrt{2}$. In addition, at $\sigma \approx 0.82$ the stability limit of Fig. 5.5, ≈ 4.45, is breached.

More complicated examples tend to produce roughly analogous results.

5.8   Changing the Viewpoint

We have, in the previous sections, taken institutional dynamics as primarily driven by the ability of a cognitive enterprise to adjust to changes in the rate of resource supply, i.e., as determined by the relation $dZ/dt = f(Z)$ in Eq. (5.20), with stochastic burden as an afterthought. Here, by contrast, we impose some particular form of stochastic stability as the principal selection pressure, recognizing that there can be many definitions of such (e.g., Appleby et al. 2008; Khasminskii 2012).

The basic stochastic differential equation now omits $f(Z)$ and is written only in terms of $dZ/dt = Zd^2F/dZ^2$, giving

$$dZ_t = (dZ/dt)_t dt + \sigma V(Z_t)dB_t =$$
$$(dS/dZ)dt + \sigma V(Z_t)dB_t =$$
$$\left(Z_t d^2F/dZ^2\right)dt + \sigma V(Z_t)dB_t \tag{5.37}$$

where $V(Z)$ is a 'volatility' function in Z, and again, dB_t is taken as Brownian noise.

Stability in order $Q(Z)$ is determined by using the Ito Chain Rule to derive dQ_t from Eq. (5.37), then by solving the nonequilibrium steady state relation $< dQ_t >= 0$ in general for $F(Z)$. This gives

$$F(Z) = \iint -\frac{\sigma^2 (V(Z))^2 \frac{d^2}{dZ^2} Q(Z)}{2Z\frac{d}{dZ}Q(Z)} dZ\, dZ + C_1 Z + C_2 \tag{5.38}$$

In second order, so that $Q(Z) = Z^2$, and with the volatility function $V(Z) = Z$,

$$F = \frac{-\sigma^2}{4}Z^2 + C_1 Z + C_2 \tag{5.39}$$

recalling that $F(Z) = -\log[h(g(Z))]g(Z)$ from Eq. (5.29).

We can bring the calculation full circle by again requiring that $Zd^2F/dZ^2 = f(Z)$. Using Eq. (5.39) and setting $f(Z) = \beta - \alpha Z$, the nonequilibrium steady state value of Z is

$$Z_{nss} = \frac{\beta}{\alpha - \sigma^2/2} \tag{5.40}$$

consonant with Eq. (5.13).

A second iteration under enforced stability extends Eq. (5.20) and further generalizes the Data Rate Theorem. We now take the basic stochastic differential equation as

$$dZ_t = \left(Z_t d^2F/dZ^2 - M(Z_t)\right)dt + \sigma V(Z_t)dB_t \tag{5.41}$$

where $M(Z)$ is again the control free energy needed to stabilize an inherent instability, *but now assumed dependent on Z.*

The expression $< dQ_t > = 0$ then leads to the relation

$$F(Z) =$$

$$\int \int \frac{-\sigma^2 (V(Z))^2 \frac{d^2}{dZ^2} Q(Z) + 2 M(Z) \frac{d}{dZ} Q(Z)}{2 Z \frac{d}{dZ} Q(Z)} dZ\, dZ$$

$$+ C_1 Z + C_2 \qquad (5.42)$$

Taking $M(Z) = M$ in second order under simple volatility, i.e., $V(Z) = Z$, Eq. (5.39) becomes

$$F = M \ln(Z) Z - MZ - \frac{\sigma^2 Z^2}{4} + C_1 Z + C_2 \qquad (5.43)$$

which is quite another matter.

Taking $M(Z) = M Z$ in second-order stability for simple volatility gives

$$F = \frac{Z^2}{2} \left(-\frac{\sigma^2}{2} + M \right) + C_1 Z + C_2 \qquad (5.44)$$

suggesting a critical value relation between $\sigma^2/2$ and M in this case.

We can again bring the calculation full circle by setting $dZ/dt = Z d^2 F/dZ^2 - M(Z) = f(Z) \geq 0$. For example, if Z is delivered according to an 'exponential' model $f(Z) = \beta - \alpha Z$, second order stability in simple volatility again gives Eq. (5.40).

Taking another perspective, if, for a 'continuous' model, the partition function sum in the second expression of Eq. (5.30) can be expressed as an integral, then $\exp[-F/g] \approx g$, and $g \approx -F/W(n, -F)$, where $W(n, -F)$ is the Lambert W-function of order n, real-valued only for $n = 0, -1$. Then a further condition for stability of 'Order Q' in Eqs. (5.43) and (5.44) is that $-F \geq -\exp[-1]$, the real-value restriction on the Lambert W-function for $n = 0, -1$.

It is also worth noting that, taking $Q(Z) = \log[Z]$ instead of Z^2 for the simple volatility models above generates the same relations as Eqs. (5.43) and (5.44), but with the sign of the σ^2 terms positive.

As Appleby et al. (2008); Khasminskii (2012), and many others indicate, the dynamics of systems that can be approximated by stochastic differential equations may be quite subtle.

5.9 Discussion

Surprisingly direct information theory approaches to system control dynamics model institutional performance under selection pressures of environmental uncertainty, delay, and 'competition', in a large sense. Beyond scientific insight, the ultimate intent is construction of new, relatively simple, but very powerful, 'worst case' statistical tools for improved, if always severely limited (Wallace 2020, 2022a), real-time control under such challenge.

Several points emerge from this study.

1. Punctuated 'phase changes' in institutional cognition and performance are to be expected, a consequence of both complex and real-valued 'Fisher zero-like' temperature analogs appearing at critical points. This work suggests that punctuated phase transition phenomena can manifest through 'inverted-U' signal transduction, by onset of thrashing/panic transitions, or by oscillations driven by response delay.
2. The multiple tunable workspaces available to the relatively slow processes of institutional cognition (Theiner et al. 2010; Wallace 2020) facilitate early recognition of 'sensory signals' representing patterns of threat or affordance. This mechanism provides enduring evolutionary advantage to social entities that can build and maintain appropriate institutional networks under persistent and acute selection pressures (e.g., Gunji et al. 2018). The proviso is that institutional doctrine, the corporate genome, permits exercise of such advantage.
3. These dynamics ultimately generalize the on/off perceptions of individual higher animal consciousness, restricted to a single 'tunable global workspace' by the necessity of containing neural dynamics within a 100 millisecond time window (e.g., Wallace 2005, 2012, 2016, 2017, 2022b; Baars 2005; Baars and Franklin 2003).
4. While stable adaptation/friction functional responses, designated as $f(Z)$ in Eq. (5.6) et seq., can be challenged by inherent limits to stochastic stability, critical 'higher constructs'—here, iterated free energies and their associated cognition rates—may be driven to instability at far lower burdens of 'noise'.
5. It appears possible to generalize control theory's Data Rate Theorem to a more comprehensive limit on overall resource rate availability, contingent on appropriate scalarization. Extension to more complicated algebraic resource structures seems possible, at the expense of some considerable mathematical complexity.
6. Following both Appleby et al. (2008) and the optimization calculation, theory suggests that a 'noise function'/'shadow price' can always be imposed on an institution as a selection pressure that will drive it to stochastic instability or other failure. Practicality is, of course, another matter, but recent experience across a number of venues—most particularly in Afghanistan, the traditional graveyard of empires—has shown that persistent adversaries may well succeed in carrying out such a program over appropriately 'evolutionary' time scales.

The emergence of these results from simplified probability models suggests that statistical tools based on this work might be useful for real-time estimate of (at least) 'worst case' institutional sensitivities to 'noise' and shadow price burdens, and for evaluation of specific multiple workspace conformations and sufficiently flexible engagement doctrines and policies when confronted by particular patterns of challenge and affordance. In addition, manipulation of basic system parameters and environmental boundary conditions materially affects sensitivity to such burdens, opportunities, and more general selection pressures.

One inference overshadows this work. Following Theiner et al. (2010), Wallace (2022b, 2020), and others, institutional cognition, in view of its many possible multiple global workspace structures and dynamics, must be recognized as far more complex than individual human consciousness. This is because consciousness is normally restricted to a single tunable workspace, ensuring rapid response to changing patterns of threat and affordance. Systems encompassing simultaneous, multiple, dynamic workspaces are qualitatively different. Consciousness is, at present, socially constructed as a most difficult scientific problem Wallace (2022b), but, it seems, is only the tip of an iceberg that must include individual humans in close synergism with their institutions, larger socioeconomies, machines, and related cultural artifacts (Theiner et al. 2010).

References

Appleby, J., X. Mao, and A. Rodkina, 2008. Stabilization and destabilization of nonlinear differential equations by noise. IEEE Transactions on Automatic Control 53:68–69.

Atlan H., and I. Cohen, 1998. Immune information, self-organization, and meaning. International Immunology 10:711–717.

Baars, B., 2005. Global workspace theory of consciousness: toward a cognitive neuroscience of human experience. Progress in Brain Research 150:45–53.

Baars, B., and S. Franklin, 2003. How conscious experience and working memory interact. Trends in Cognitive Science 7:166–172.

Brown, R., 1992. Out of line. Royal Institute Proceedings 64:207–243.

Cayron, C., 2006. Groupoid of orientational variants. Acta Crystalographica Section A A62:21040.

Champagnat, N., R. Ferriere, and S. Meleard, 2006, Unifying evolutionary dynamics: from individual stochastic process to macroscopic models. Theoretical Population Biology 69:297–321.

Cover, T., and J. Thomas, 2006. *Elements of Information Theory*, 2nd ed. New York: Wiley.

de Groot, S., and P. Mazur, 1984. *Nonequilibrium Thermodynamics*. New York: Dover.

Dembo, A., and O. Zeitouni, 1998. *Large Deviations and Applications*, 2nd ed. New York: Springer.

Diamond, D., A. Campbell, C. Park, J. Halonen, and P. Zoladz, 2007. The temporal dynamics model of emotional memory processing neural plasticity. https://doi.org/10.1155/2007/60803.

Dolan, B., W. Janke, D. Johnston, and M. Stathakopoulos, 2001. Thin Fisher zeros. Journal of Physics A 34:6211–6223.

Dretske, F., 1994. The explanatory role of information. Philosophical Transactions of the Royal Society A 349:59–70.

Feynman, R. 2000. *Lectures on Computation*. New York: Westview Press.

Fisher, M., 1965. *Lectures in Theoretical Physics*, Vol. 7. Boulder: University of Colorado Press.

Golubitsky, M., and I. Stewart, 2006. Nonlinear dynamics and networks: the groupoid formalism. Bulletin of the American Mathematical Society 43:305–364.

Gunji, Y., H. Murakami, T. Tomaru, and V. Basios, 2018. Inverse Baysian inference in swarming behaviour of soldier crabs. Philosophical Transactions A 376:21070370.

Jin, H., Z. Hu, and X. Zhou, 2008. A convex stochastic optimization problem arising from portfolio selection. Mathematical Finance 18:171–183.

Khasminskii, R., 2012. *Stochastic Stability of Differential Equations*, 2nd ed. New York: Springer.

Khinchin, A., 1957. *Mathematical Foundations of Information Theory*. New York: Dover.

Laidler, K., 1987. *Chemical Kinetics*, 3rd ed. New York: Harper and Row.

Nair, G., F. Fagnani, S. Zampieri, and R. Evans, 2007. Feedback control under data rate constraints: an overview. Proceedings of the IEEEE 95:108–137.

Nocedal, J., and S. Wright, 2006. *Numerical Optimization*, 2nd ed. New York: Springer.

Osinga, F., 2007. *Science, Strategy and War: The Strategic Theory of John Boyd*. London: Routledge.

Pettini, M., 2007. *Geometry and Topology in Hamiltonian Dynamics and Statistical Mechanics*. New York: Springer.

Protter, P., 2005. *Stochastic Integration and Differential Equations: A New Approach*, 2nd ed. New York: Springer.

Robinson, S., 1993. Shadow prices for measures of effectiveness II: general model. Operations Research 41:536–548.

Ruelle, D., 1964. Cluster property of the correlation functions of classical gases. Reviews of Modern Physics April:580–584.

Shomorony, I., and A. Avestimehr, 2012. Is Gaussian noise the worst-case additive noise in wireless networks? In *IEEE International Symposium on Information Theory Proceedings 2012*, 214–218. https://doi.org/10.1109/ISIT.2012.6283743.

Theiner, G., C. Allen, and R. Goldstone, 2010. Recognizing group cognition. Cognitive Systems Research 11:378–395.

Wallace, R., 2005. *Consciousness: A Mathematical Treatment of the Global Neuronal Workspace Model*. New York: Springer.

Wallace, R., 2012. Consciousness, crosstalk, and the mereological fallacy: an evolutionary perspective. Physics of Life Reviews 9:426–453.

Wallace, R., 2015. *An Ecosystem Approach to Economic Stabilization*. London: Routledge Press.

Wallace, R., 2016. Environmental induction of neurodevelopmental disorders. Bulletin of Mathematical Biology 78:2408–2426.

Wallace, R., 2017. *Computational Psychiatry: A Systems Biology Approach to the Epigenetics of Mental Disorders*. New York: Springer.

Wallace, R., 2020. *Cognitive Dynamics on Clausewitz Landscapes: The Control and Directed Evolution of Organized Conflict*. New York: Springer.

Wallace, R., 2022a. *Essays on Strategy and Public Health: The Systematic Reconfiguration of Power Relations*. New York: Springer.

Wallace, R., 2022b. *Consciousness, Cognition and Crosstalk: The Evolutionary Exaptation of Nonergodic Groupoid Symmetry-Breaking*. New York: Springer.

Wallace, R., and Fullilove, 2014. State policy and the political economy of criminal enterprise: mass incarceration and persistent organized hyperviolence in the USA. Structural Change and Economic Dynamics 31:17–31.

Weinstein, A., 1996. Groupoids: unifying internal and external symmetry. Notices of the American Mathematical Association. 43:744–752.

Chapter 6
On 'Speciation': Fragment Size in Information System Phase Transitions

6.1 Introduction

The previous chapters explored possible applications of a relatively simple dynamic model of cognition based on a first-order approximation similar to the Onsager treatment of nonequilibrium thermodynamics. This was constrained by the inherent directed homotopy/groupoid foundation of information sources, i.e., no microreversibility, and hence no 'Onsager reciprocal relations'. Some piecemeal extensions of the formalism were explored, but here we more systematically outline how phase transition models in both 'cognitive' and 'heritage' information systems might be expanded to include more complicated phase transition dynamics.

6.2 'Simple' Phase Transition

Recall the argument regarding information dynamics for nonergodic systems, where information source uncertainty is path dependent, i.e., each high probability path has its own limiting value of source uncertainty, not, however, given as a 'Shannon entropy' (Khinchin 1957).

For each high probability path j it is then possible to write a Boltzmann pseudoprobability as

$$P_j = \frac{\exp[-H_j/g(Z)]}{\sum_k \exp[-H_k/g(Z)]} \tag{6.1}$$

H_k is the source uncertainty of the high probability path k and the sum—or generalized integral—is over all high probability paths. $g(Z)$ is a scalar 'cognitive temperature' analog depending on an essential resource rate Z, assumed here to also be a scalar. See Wallace (2020) for address of a generalization of Z to

R. Wallace, *Essays on the Extended Evolutionary Synthesis*, SpringerBriefs in Evolutionary Biology, https://doi.org/10.1007/978-3-031-29879-0_6

multidimensional form. Z may index available internal or external information bandwidths, rate of metabolic free energy in a physiological system, rate of 'materiel' and personnel supply in organized conflict, or some synergism of these that, like a principle component analysis, projects down to one dimension but accounts for a substantial fraction of overall variance.

We then define a free energy Morse Function F (Pettini 2007) in terms of the 'partition function' denominator of Eq. (6.1),

$$\exp[-F/g(Z)] = \sum_k \exp[-H_k/g(Z)] \equiv h(g(Z))$$

$$F(Z) = -\log[h(g(Z))]g(Z)$$

$$g(Z) = -\frac{F(Z)}{RootOf\left(e^X - h\left(-\frac{F(Z)}{X}\right)\right)} \tag{6.2}$$

where X is a dummy variate whose solution gives the desired expression. Taking $h(g(Z)) = g(Z)$ shows the RootOf construct generalizes the Lambert W-function of order n that satisfies the relation $W(n, x)\exp[W(n, x)] = x$. In that case, $g(Z) = -F(Z)/W(n, -F(Z))$.

A central argument has been that the RootOf relation for $g(Z)$ may have complex solutions, representing phase transitions analogous to those of 'Fisher zeros' in physical systems (e.g., Dolan et al. 2001 and references therein). Here, we abduct a different perspective, based on the Kadanoff/Wilson view of physical phase transition.

6.3 Phase Transitions in Networks of Information-Exchange Modules

We follow closely the arguments Wallace (2022, Ch. 4). Consider an interlinked network of information-exchange modules that must act individually under some—perhaps most—circumstances, but coherently under others. Possible examples range from the intracellular regulation of protein folding or gene expression through institutional response in armed conflict. How do such linkages come about?

Previous sections abducted results from nonequilibrium thermodynamics to characterize the dynamics of cognitive process. Here, we abduct the Kadanoff renormalization treatment of physical phase transitions (e.g., Wilson 1971; Wallace 2022), applying it to a reduced version of the iterated 'free energy' Morse Function of Eq. (6.2).

We will be primarily concerned with internal system bandwidth—the strength of crosstalk between modules—envisioning a number of internal cognitive submodules as connected into a topologically identifiable network having a variable average number of fixed-strength crosstalk linkages between components. The mutual information measure of crosstalk can continuously change, and it becomes then possible to conduct a parameterized renormalization in a now-standard manner (Wilson 1971; Wallace 2022).

The internal modular network linked by information exchange has a topology depending on the magnitude of interaction. We define an interaction parameter, a real number $\omega > 0$, and examine structures characterized in terms of linkages set to zero if crosstalk is less than ω, and renormalized to 1 if greater than or equal to ω. Each ω defines, in turn, a network 'giant component' (Spenser 2010; Newman 2010), linked by information exchange greater than or equal to it.

Recall the standard result of Fig. 1.3 that, in a random network of N nodes, a 'giant component' emerges when the probability of lineage P exceeds a threshold. The fraction of nodes in that component, $\mathcal{N}$, and the related threshold probabilities, are given as

$$\mathcal{N} = \frac{1}{NP} \left(W(0, -NP \exp[-NP]) + NP \right)$$

$$P = \frac{\log(-\frac{1}{\mathcal{N}-1})}{\mathcal{N}N} \propto \frac{1}{N} \tag{6.3}$$

where $W(0, x)$ is the Lambert W-function of order 0, remembering that this is real-valued only for $x > -\exp[-1]$, so that there will be a threshold in the index NP, topping out at the full network. The second part of Eq. (6.3) is consonant with speciation arguments related to percolation across rough high dimensional 'fitness landscapes' (e.g., Gavrilets 2010).

It is easy to see that, for a 'stars-of-stars-of-stars' network—highly nonrandom indeed—there will be no threshold, although the rate of propagation may be slow. Thus network topology is important per se.

The central trick is to invert the argument: a given topology of interacting submodules making up a giant component will, in turn, define some critical value ω_C such that network elements interacting by information exchange at a rate less than that value will be excluded from that component, will be locked out.

ω is a tunable, syntactically dependent, detection limit depending on the instantaneous topology of the giant component of linked cognitive submodules defining, by that linkage, a 'global broadcast'.

For 'slow' systems—protein folding regulation, immune response, gene expression, institutional process—there can be many such 'global workspace' spotlights acting simultaneously. Such multiple global broadcasts, indexed by the set $\Omega = \{\omega_1, \omega_2, \ldots\}$, can—if allowed to function by doctrine or other internal censors—lessen the likelihood of inattentional blindness to critical signals, both internal and external. The immune system, for example, engages simultaneously in pathogen and malignancy attack, neuroimmuno dialog, and routine tissue maintenance.

Assuming it possible to scalarize the set Ω in something of the manner of Z above, we project down to a single, real-value ω, and model the dynamics of a multiple tunable workspace system using the Kadanoff formalism.

Recall the definition of the iterated free energy F from Eq. (6.2), now focused within and characterized by the reduced scalar ω. The approach relies on a 'length' r on the network of internal interacting information sources. r will be more fully defined below. We follow the classic renormalization methodology of Wilson (1971) as described in Wallace (2022), although there is no unique renormalization symmetry.

The central idea is to invoke a 'clumping' transformation under an 'external field strength' that can be, in the limit, set to zero. For clumps of size R, given a field of strength J,

$$F[\omega(R), J(R)] = \mathscr{F}(R)F[\omega(1), J(1)]$$

$$\chi[\omega(R), J(R)] = \frac{\chi[\omega(1), J(1)]}{R} \tag{6.4}$$

χ represents a correlation length across the linked information sources.

$\mathscr{F}(R)$ is a 'biological' renormalization relation that can take such forms as R^δ, $m\log(R) + 1$, $\exp[m(R - 1)/R]$, and so on, so long as $\mathscr{F}(1) = 1$ and is otherwise monotonic increasing. Physical theory is restricted to $\mathscr{F}(R) = R^3$.

Surprisingly, after some tedious algebra, the standard Wilson (1971) renormalization phase transition calculation drops right out for the extended relations, summarized in the chapter Mathematical Appendix, after Wallace (2005, 2022).

What is the metric r? Again, we follow Wallace (2022).

First, impose a topology on the system of interacting information sources such that, near a particular 'language' A associated with some source uncertainty measure H, there is an open set U of closely similar languages $\hat{A}$ such that the set $A, \hat{A} \in U$.

Since the information sources are sufficiently similar, for all pairs of languages $A, \hat{A}$ in U it is possible to

- Create an embedding alphabet which includes all symbols allowed to both.
- Define an information-theoretic distortion measure in the extended joint alphabet between any high probability (i.e, properly grammatical and syntactical) paths in A and $\hat{A}$, written as $d(Ax, \hat{A}x)$. The different languages do not interact in this approximation.
- Define the metric on U as

$$r(A, \hat{A}) \equiv |\int_{A,\hat{A}} d(Ax, \hat{A}x) - \int_{A,A} d(Ax, A\hat{x})| \tag{6.5}$$

where Ax and $\hat{A}x$ are paths in the languages $A, \hat{A}$ respectively, d is the distortion measure, and the second term is a 'self-distance' for the language A such that $r(A, A) = 0, r(A, \hat{A}) > 0, A \neq \hat{A}$.

More detailed consideration shows this version of r is sufficient to the task (Glazebrook and Wallace 2009).

Extension of the Wilson technique seems straightforward. However, since the dynamics of the embedded condition are so highly variable, there will be no unique solution, although there may well be equivalence classes of solutions, defining yet more goupoids in the sense of Tateishi et al. (2013).

Groupoids may appear at a far more fundamental level: Wilson's renormalization semigroup, in the cognitive circumstance of discrete equivalence classes of developmental pathways, might well require generalization as a renormalization semigroupoid, e.g., a disjoint union of different renormalization semigroups across a nested or otherwise linked set of information sources and/or iterated free energy constructs dual to cognitive modules.

In effect, we have studied equivalence classes of directed homotopy developmental paths associated with nonergodic cognitive systems defined in terms of single-path source uncertainties. These require imposition of structure in terms of the metric r of Eq. (6.5), leading to groupoid symmetry-breaking transitions driven by changes in the temperature analog $g(Z)$.

There can be an intermediate case under circumstances in which the standard ergodic decomposition of a stationary process is both reasonable and computable—no small constraint. Then there is an obvious natural directed homotopy partition in terms of the transitive components of the path-equivalence class groupoid. This decomposition seems equivalent to, and maps on, the ergodic decomposition of the overall stationary cognitive process. It then becomes possible to define a constant source uncertainty on each transitive subcomponent, fully indexed by the embedding groupoid.

That is, each ergodic/transitive groupoid component of the ergodic decomposition recovers a constant value of the source uncertainty dual to cognition, presumably given by standard 'Shannon entropy' expression. Since it is possible to view the components themselves as constituting single paths in an appropriate quotient space, leading to the previous 'nonergodic' developments.

A complication emerges through imposition of a double symmetry involving metric r-defined equivalence classes on this quotient space. That is, there are different possible strategies for any two teams playing the same game. In sum, however, groupoid symmetry-breaking in the iterated free energy construct of Eqs. (6.2) will still be driven by changes in $g(Z)$ and/or ω.

We can, under certain restrictions, estimate fragment size after a phase transition event.

Assume a critical value ω as ω_C and define $\kappa \equiv (\omega_C - \omega)/\omega_C$. Next, assume $\mathscr{F}(R) = R^\delta$, and substitute into Eq. (6.4), taking the limit $J \to 0$. Following Wallace (2015, Section 7.4) gives, in first order near ω_C,

$$F = \kappa^{\delta/y} F_0$$
$$\chi = \kappa^{-1/y} \chi_0 \tag{6.6}$$

where y, F_0, and χ_0 are constants coming from the series expansion.

Next, following Zurek (1985, 1996), assume the rate of change of κ remains constant, with $|d\kappa/dt| \equiv 1/\tau_\omega$. Argument from physical theory suggests there is a characteristic time constant for the phase transition, $\tau = \tau_0/\kappa$ such that, if changes in κ take place on a time scale longer than τ for any given κ, the correlation length $\chi = \chi_0 \kappa^{-s}$, $s = 1/y$, will be in equilibrium with internal changes, and result in a very large fragment in R-space.

Zurek next argues that a critical freeze-out time $\hat{t}$ occurs at a system time $\hat{t} = \chi/|d\chi/dt|$ such that $\hat{t} = \tau$. Evaluating $d\chi/dt$, remembering the definition $d\kappa/dt = 1/\tau_\omega$, then

$$\frac{\chi}{|d\chi/dt|} = \frac{\kappa\tau_\omega}{s} = \frac{\tau_0}{\kappa}$$

$$\kappa = \sqrt{s\tau_0/\tau_\omega} \qquad (6.7)$$

Substitution of this value for κ into the expression for correlation length χ gives the expected size of fragments in R-space, say $d(\hat{t})$, as

$$d \approx \chi_0 (\frac{\tau_\omega}{s\tau_0})^{s/2} \qquad (6.8)$$

Here, $s = 1/y > 0$.

In consequence, the more rapidly ω approaches the critical value ω_C, the smaller is τ_ω, and the smaller and more numerous are the R-space fragments, more likely to risk extinction under selection pressures, or in the case of something like the break-up of a large illegal enterprise, more likely to engage in hyperviolent evolutionary market competition. Again, see Wallace (2015) for a more complete analysis.

6.4   Discussion

The fragmentation dynamics explored here extend earlier Fisher Zero 'phase transition' perspectives on punctuated changes in information systems. Group/groupoid 'symmetry-breaking' implies that 'crystalline domains' of lower symmetry will emerge as possible states of the system. The Zurek calculation permits estimation of the size of those domains. Depending on the nature of the embedding selection pressures, in particular for institutional function, it seems probable that larger domains will be more likely to survive. In any event, subsequent evolutionary processes will act on the surviving fragmentary domains, leading to 'speciation', in a large sense.

6.5   Mathematical Appendix: 'Biological' Renormalizations

We adapt the renormalization scheme of Wallace (2005, 2022), focused on a stationary, ergodic, information source H, to the generalized free energy associated with nonergodic cognition.

Equation (6.4) states that the information source and the correlation length, the degree of coherence on the underlying network, scale under renormalization clustering in chunks of size R as

$$F[\omega(R), J(R)] = \mathscr{F}(R)F[\omega(1), J(1)]$$

$$\chi[\omega(R), J(R)]R = \chi[\omega(1), J(1)]$$

with $\mathscr{F}(1) = 1$.

Differentiating these two equations with respect to R, so that the right hand sides are zero, and solving for $d\omega(R)/dR$ and $dJ(R)/dR$ gives, after some manipulation,

$$d\omega_R/dR = u_1 d\log(\mathscr{F})/dR + u_2/R$$

$$dJ_R/dR = v_1 J_R d\log(\mathscr{F})/dR + \frac{v_2}{R}J_R \qquad (6.9)$$

The $u_i, v_i, i = 1, 2$ are functions of $\omega(R), J(R)$, but not explicitly of R itself.

We expand these equations about the critical value $\omega_R = \omega_C$ and about $J_R = 0$, obtaining

$$d\omega_R/dR = (\omega_R - \omega_C)yd\log(\mathscr{F})/dR + (\omega_R - \omega_C)z/R$$

$$dJ_R/dR = wJ_R d\log(\mathscr{F})/dR + xJ_R/R \qquad (6.10)$$

The terms $y = du_1/d\omega_R|_{\omega_R=\omega_C}, z = du_2/d\omega_R|_{\omega_R=\omega_C}, w = v_1(\omega_C, 0), x = v_2(\omega_C, 0)$ are constants.

Solving the first of these equations gives

$$\omega_R = \omega_C + (\omega - \omega_C)R^z \mathscr{F}(R)^y \qquad (6.11)$$

again remembering that $\omega_1 = \omega, J_1 = J, \mathscr{F}(1) = 1$.

Wilson's (1971) essential trick is to iterate on this relation, which is supposed to converge rapidly near the critical point, assuming that for ω_R near ω_C, we have

$$\omega_C/2 \approx \omega_C + (\omega - \omega_C)R^z \mathscr{F}(R)^y \qquad (6.12)$$

We iterate in two steps, first solving this for $\mathscr{F}(R)$ in terms of known values, and then solving for R, finding a value R_C that we then substitute into the first of Eq. (6.4) to obtain an expression for $F[\omega, 0]$ in terms of known functions and parameter values.

The first step gives the general result

$$\mathscr{F}(R_C) \approx \frac{[\omega_C/(\omega_C - \omega)]^{1/y}}{2^{1/y}R_C^{z/y}} \qquad (6.13)$$

Solving this for R_C and substituting into the first expression of Eq. (6.10) gives, as a first iteration of a far more general procedure, the result

$$F[\omega, 0] \approx \frac{F[\omega_C/2, 0]}{\mathscr{F}(R_C)} = \frac{F_0}{\mathscr{F}(R_C)}$$

$$\chi(\omega, 0) \approx \chi(\omega_C/2, 0)R_C = \chi_0 R_C \qquad (6.14)$$

giving the essential relationships.

Note that a power law of the form $\mathscr{F}(R) = R^m$, $m = 3$, which is the direct physical analog, may not be biologically reasonable, since it says that 'language richness', in a general sense, can grow very rapidly as a function of increased network size. Such rapid growth is simply not observed in cognitive process.

Taking the biologically realistic example of non-integral 'fractal' exponential growth,

$$\mathscr{F}(R) = R^\delta \qquad (6.15)$$

where $\delta > 0$ is a real number which may be quite small, equation we can be solve for R_C, obtaining

$$R_C = \frac{[\omega_C/(\omega_C - \omega)]^{[1/(\delta y + z)]}}{2^{1/(\delta y + z)}} \qquad (6.16)$$

for ω near ω_C. Note that, for a given value of y, one might characterize the relation $\alpha \equiv \delta y + z = \text{constant}$ as a 'tunable universality class relation' in the sense of Albert and Barabasi (2002).

Substituting this value for R_C back gives a complex expression for F, having three parameters: δ, y, z.

A more biologically interesting choice for $\mathscr{F}(R)$ is a logarithmic curve that 'tops out', for example

$$\mathscr{F}(R) = m \log(R) + 1 \qquad (6.17)$$

Again $\mathscr{F}(1) = 1$.

Using a computer algebra program to solve for R_C gives

$$R_C = [\frac{Q}{W[0, Q \exp(z/my)]}]^{y/z} \qquad (6.18)$$

where

$$Q \equiv (z/my)2^{-1/y}[\omega_C/(\omega_C - \omega)]^{1/y}$$

Again, $W(n, x)$ is the Lambert W-function of order n.

An asymptotic relation for $\mathscr{F}(R)$ would be of particular biological interest, implying that 'language richness' increases to a limiting value with population growth. Taking

$$\mathscr{F}(R) = \exp[m(R-1)/R] \tag{6.19}$$

gives a system which begins at 1 when $R = 1$, and approaches the asymptotic limit $\exp(m)$ as $R \to \infty$. Computer algebra finds

$$R_C = \frac{my/z}{W[0, A]} \tag{6.20}$$

where

$$A \equiv (my/z) \exp(my/z)[2^{1/y}[\omega_C/(\omega_C - \omega)]^{-1/y}]^{y/z}$$

These developments suggest the possibility of taking the theory significantly beyond arguments by abduction from simple physical models.

References

Albert, R., A. Barabasi. 2002. Statistical mechanics of complex networks. Reviews of Modern Physics 74:47–97.

Dolan, B., W. Janke, D. Johnston, and M. Stathakopoulos. 2001. Thin Fisher zeros. Journal of Physics A 34:6211–6223.

Gavrilets, S. 2010. High-dimensional fitness landscapes and speciation. In *Evolution: The Extended Synthesis*, ed. Massimo Pigliucci and Gerd B. Müller. Cambridge: MIT Press.

Glazebrook, J.F., and Wallace, R. 2009. Rate distortion manifolds as model spaces for cognitive information. Informatica 33:309345.

Khinchin, A. 1957. *Mathematical Foundations of Information Theory*. New York: Dover.

Newman, M. 2010. *Networks: An Introduction*. New York: Oxford University Press.

Pettini, M. 2007. *Geometry and Topology in Hamiltonian Dynamics and Statistical Mechanics*. New York: Springer.

Spenser, J. 2010. The giant component: a golden anniversary. Notices of the American Mathematical Society 57:720–724.

Tateishi, A., R. Hanel, and S. Thurner. 2013. The tranformation groupoid structure of the q-Gaussian family. Physics Letters A 377:1804–1809.

Wallace, R. 2005. *Consciousness: A Mathematical Treatment of the Global Neuronal Workspace Model*. New York: Springer.

Wallace, R. 2015. *An Ecosystem Approach to Economic Stabilization*. London: Routledge Press.

Wallace, R. 2020. How AI founders on adversarial landscapes of fog and friction. Journal of Defense Modeling and Simulation. https://doi.org/10.1177/1548512920962227

Wallace, R. 2021. Toward a formal theory of embodied cognition. BioSystems 202:104356.

Wallace, R. 2022. *Consciousness, Cognition and Crosstalk: The Evolutionary Exaptation of Nonergodic Groupoid Symmetry-Breaking*. New York: Springer.

Wilson K. 1971. Renormalization group and critical phenomena. I Renormalization group and the Kadanoff scaling picture. Physics Reviews B 4:317483.

Zurek, W. 1985. Cosmological experiments in superfluid helium? Nature 317:505–508.

Zurek, W. 1996. The shards of broken symmetry. Nature 382:296–298.

Chapter 7
Adapting Cognition Models to Biomolecular Condensate Dynamics

The prebiotic organization of chemicals into compartmentalized ensembles is an essential step to understand the transition from inert molecules to living matter. Compartmentalization is indeed a central property of living systems... ...[D]ifferent compartments could have co-emerged, competed for the same resources, or collaborated to 'survive' until one population would have acquired a selective advantage making it thrive at the expense of the other populations.

— Martin and Douliez (2021)

Living systems are cognitive systems, and living as a process is a process of cognition. This statement is valid for all organisms, with and without a nervous system.

— (Maturana and Varela 1980, p. 13)

7.1 Introduction

The seminal paper by Peeples and Rosen (2021) characterizes biomolecular condensates as follows:

Biomolecular condensates concentrate proteins and RNA molecules without a surrounding membrane. Condensates appear in a wide range of biological contexts, including mRNA storage/degradation, T-cell activation, and ribosome biogenesis. Many condensates appear to form through liquid-liquid phase-separation (LLPS), in which oligomerization mediated by multivalent interactions lowers the solubility of proteins and/or nucleic acids sufficiently to form a second phase. Where measured, the degree of concentration in this phase ranges from 2- to 150-fold for different constituents. Scaffold-like molecules, those that contribute more strongly to formation of the condensate, are typically the most highly concentrated components...

Condensates provide potential mechanisms by which cells can regulate biochemical processes temporally and spatially. Because condensates concentrate enzymes and their potential substrates, the compartments are often invoked as accelerating biochemical reactions.

R. Wallace, *Essays on the Extended Evolutionary Synthesis*, SpringerBriefs in Evolutionary Biology, https://doi.org/10.1007/978-3-031-29879-0_7

Their in vitro model of the phenomenon, focusing on scaffold-like infrastructure, was found to increase a particular reaction rate by a factor of 36, leading to the striking observations of Sang et al. (2022) that phase-separated biomolecular condensates can potentate kinase signaling, related to Alzheimer's-associated phosphorylation events.

Similarly, Lyon et al. (2021) write

> Biomolecular condensates are found throughout eukaryotic cells, in the nucleus, cytoplasm and on various membranes. They are also found across the biological spectrum, organizing molecules that act in processes ranging from RNA metabolism to signaling to gene regulation... Studies have revealed the central role of multivalent interactions in driving condensate formation. Many instantiations of multivalency have been described in condensates, involving folded protein domains, intrinsically disordered regions (IDR) nucleic acids, and chromatin. For IDRs, a 'molecular grammar' model has emerged, wherein the abundance and patterning of certain amino acids within the sequence influences both the drive to form and the physical properties of the condensates (Wang et al. 2018; Martin et al. 2020; Lin et al. 2018; Vernon et al. 2018)...

Here, specifically focusing on the 'grammar' of these reactions, we translate ideas of chemical reaction rate into a recently-developed dialect of information-theoretic based discussions of cognition rate, reflecting more exactly the injunction of Maturana and Varela (1980) above. To do this we must first build a theoretical scaffolding upon which to hang our subsequent discussions. This is not entirely simple, if surprisingly straightforward. We first review something of Wallace (2022).

7.2   Resources

At least three resource streams are required by a cognitive entity. The first is measured by the rate $\mathscr{C}$ at which information can be transmitted between elements within the entity, determined as an information channel capacity (Cover and Thomas 2006). The second stream is sensory information regarding the embedding environment, available at a rate $\mathscr{Q}$. The third regards material resources, including metabolic free energy—in a large sense—available at a rate $\mathscr{M}$. These rates will usually change in time, leading to the necessity of developing a dynamic theory under highly nonequilibrium circumstances. This is not a trivial matter, and requires some considerable methodological development.

These resource streams must interact, characterized by a 3 by 3 matrix analogous to, but not the same as, an ordinary correlation matrix, here written as $\mathbf{Z}$.

An n-dimensional square matrix $\mathbf{M}$ has n scalar invariants $r_1, \ldots, r_n$ defined by the characteristic equation for $\mathbf{M}$:

$$p(\gamma) = \det[\mathbf{M} - \gamma \mathbf{I}] =$$
$$(-1)^n \gamma^n + (-1)^{n-1} r_1 \gamma^{n-1} + (-1)^{n-2} r_2 \gamma^{n-2} - \ldots - r_{n-1}\gamma + r_n \quad (7.1)$$

I is the n-dimensional identity matrix, det the determinant, and γ a real-valued parameter. The first invariant, r_1, is usually taken as the matrix trace, and the last, r_n, as the determinant.

These scalar invariants make it possible to project the full matrix down onto a single scalar index $M = M(r_1, \ldots, r_n)$ retaining—under some circumstances—much of the basic structure, analogous to conducting a principal component analysis, which does much the same thing for a correlation matrix (e.g., Jolliffe 2002). Wallace (2021a) provides an example in which two such indices are necessary, leading to serious mathematical complexities.

Here, the simplest such index might be $Z = \mathscr{C} \times \mathscr{Q} \times \mathscr{M}$. However, there will almost always be important cross-interactions between different resource streams, requiring a more complete analysis, that is, one based Eq. (7.1) that takes crossterms explicitly into account. That is the assumption we make here, a most central and difficult scientific question.

Clever scalarization—when appropriate—enables approximate reduction to a one-dimensional system. Again, expansion of Z into vector form is possible, but leads to difficult multidimensional dynamic equations (Wallace 2021a).

7.3 Cognition

Here, we lift the ergodic restriction on information sources (Cover and Thomas 2006). Only in the case that cross-sectional and longitudinal means are the same can information source uncertainty be expressed as a conventional Shannon 'entropy' (Khinchin 1957). We do require that source uncertainties converge for sufficiently long paths, not that they fit some particular functional form. It is the values of those uncertainties that will be of concern, not their functional expressions. We will study what might be called Adiabatically Piecewise Stationary (APS) systems, in the sense of the Born-Oppenheimer approximation for molecular systems that assume nuclear motions are so slow in comparison with electron dynamics that they can be effectively separated, at least on appropriately chosen trajectory 'pieces' that may characterize the various phase transitions available to such systems. Extension of this work to nonstationary circumstances remains to be done. Specifically, between phase transitions, we assume that the system changes slowly enough so that the asymptotic limit theorems of information and control theories can be invoked.

We carry out this approximation via a fairly standard Morse Function iteration (e.g., Pettini 2007).

Our systems of interest are composed of cognitive submodules that engage in crosstalk. At every scale and level of organization all such submodules are constrained by both their own internals and developmental paths and by the persistent regularities of the embedding environment, including the cognitive intent of adversaries and the regularities of 'grammar' and 'syntax' imposed by embedding evolutionary and environmental pressures.

Further, there are structured uncertainties imposed by the large deviations possible within that environment, again including the behaviors of adversaries who may be constrained by quite different developmental trajectories and 'punctuated equilibrium' evolutionary transitions.

Recapitulating somewhat the arguments of Wallace (2018, 2020a), the Morse Function construction assumes a number of interacting factors:

1. As Atlan and Cohen (1998) argue, cognition requires choice that reduces uncertainty. Such reduction in uncertainty directly implies the existence of an information source 'dual' to that cognition at each scale and level of organization. The argument is unambiguous and sufficiently compelling. Again, see Wallace (2012, Sec. 4) for a more complete explication.
2. Cognitive physiological processes, like the immune and gene expression systems, are highly regulated, in the same sense that 'the stream of consciousness' flows between cultural and social 'riverbanks'. That is, a cognitive information source X_i is generally paired with a regulatory information source X^i.
3. Environments (in a large sense) impose temporal event sequences of very high probability: night follows day, hot seasons follow cold, wet season follows dry, and so on. Thus environments impose their own 'meaningful statements' onto entities and interactions embedded within them via an information source V.
4. 'Large deviations', following Champagnat et al. (2006) and Dembo and Zeitouni (1998), also involve sets of high probability developmental pathways, often governed by 'entropy'-like laws that imply the existence of yet one more information source L_D.

Full system dynamics must then be characterized by a joint—*path dependent*—nonergodic information source uncertainty

$$H(\{X_i,\ X^i\}, V, L_D) \tag{7.2}$$

Individual dynamic paths of sufficient length can then be assigned a value for that joint source uncertainty, denoted by $H(x)$ for a path x.

This 'fundamental representation' is now defined by individual dynamic path values of source uncertainty and not represented as an 'entropy' function defined for all high-probability paths by an underlying probability distribution (Khinchin 1957). That is, *each path has it's own H-value*, but it is not in terms of a 'Shannon entropy' across an underlying probability distribution for all paths.

Again, the set $\{X_i,\ X^i\}$ includes the internal interactive cognitive dual information sources of the system of interest and their associated regulators, V is taken as the information source of the embedding environment that may include the actions and intents of adversaries/symbionts, as well as 'weather'. L_D is the information source of the associated large deviations possible to the system.

As above, we project the matrix of essential resources and their interactions onto a scalar rate index Z, according to the argument following Eq. (7.3). This may not always be possible, leading to significant multidimensional complexities (Wallace 2021a).

An essential insight—following from the properties of nonergodic information sources—is that the underlying equivalence classes of developmental-behavioral-dynamic system paths used to define groupoid symmetries can now be defined fully in terms of the magnitude of individual path source uncertainties *of individual dynamic paths* $H(x_j)$ such that $x_j = \{x_j^0, x_j^1, \ldots x_j^n, \ldots\}$ at times $m = 0, 1, 2, \ldots n \to \infty$ alone. See Khinchin (1957) for details of the nonergodic limit argument. Individual paths of sufficient length have associated source uncertainty scalar values, even if not calculated as standard Shannon 'entropies' across some probability distribution.

One can envision the equivalence classes of behavioral/developmental paths as defined by the 'game' the organism is playing: growing from inception, foraging for food or habitat, evading predation, wound healing, mating/reproducing, and so on. Paths within each 'game' are taken as equivalent. By the arguments of the Chapter Mathematical Appendix, this division of developmental/behavioral paths defines a groupoid. From a human perspective, sets of behavioral paths associated with baseball, football, soccer, rugby, tennis, and so on, are easily discernible and placed in appropriate equivalence classes.

Recall, as well, the conundrum of the ergodic decomposition of nonergodic information sources. It is formally possible to express a nonergodic source as the composition of a sufficient number of ergodic sources, much as it is possible to reduce planetary orbits to a Fourier sum of circular epicycles, obscuring the basic dynamics. Hoyrup (2013) discusses the problem further, finding that ergodic decompositions are not necessarily computable. Here, we finesse the matter by focusing only on the values of the source uncertainties associated with dynamic paths.

7.4 Phase Transitions I: Fisher Zeros

The next step is to build an *iterated* 'free energy' Morse Function (Pettini 2007) from a Boltzmann pseudoprobability, based on enumeration of high probability developmental pathways x_j, $j = 1, 2, \ldots$ available to the system—each having an individual joint uncertainty $H(x_j) \equiv H_j$ so that

$$P_j = \frac{\exp[-H_j/g(Z)]}{\sum_k \exp[-H_k/g(Z)]} \tag{7.3}$$

where H_j is the source uncertainty of the high probability path j, which—again—we do not assume to be given as a 'Shannon entropy' since we are no longer restricted to ergodic sources.

This step should be recognized as a version of the standard 'free energy' in statistical physics as constructed from a partition function (Landau and Lifshitz 2007), using Feynman's (2000) central insight that 'information' can, itself, be viewed as a form of free energy. Hence the iteration.

The essential point at this stage of the argument is a generalization of the fundamental assumption behind the Shannon-McMillan Theorem of information theory (Khinchin 1957). That is, in the limit of 'infinite length', *it remains possible to divide the full set of individual dynamic paths into two distinct equivalence classes,* a small set of high probability paths consonant with some underlying 'grammar' and 'syntax' that 'make sense' within the venue of the organism/system and its environment, and a much larger set of paths of vanishingly low probability not so consonant, a set of measure zero. This is a somewhat subtle point. Observational characterization of such 'grammar' and 'syntax' is not trivial, and in the case of the 'Genetic Code' required much empirical effort (e.g., Marshall 2014).

We thus infer a more general principle.

The temperature-analog characterizing the system, written as $g(Z)$ in Eq. (7.5), can be calculated via a first-order Onsager nonequilibrium thermodynamic approximation built from the partition function, i.e., the denominator of Eq. (7.5) (de Groot and Mazur 1984).

We define the 'iterated free energy' Morse Function F as

$$\exp[-F/g(Z)] \equiv \sum_k \exp[-H_k/g(Z)] \equiv h(g(Z))$$

$$F(Z) = -\log[h(g(Z))]g(Z)$$

$$g(Z) = -\frac{F(Z)}{RootOf\left(e^Q - h\left(-\frac{F(Z)}{Q}\right)\right)} \tag{7.4}$$

where Q is a dummy variate representing the solution of the RootOf construct.

The sum is over all possible high probability developmental paths of the system, again, those consistent with an underlying grammar and syntax, and the last expression is a generalized solution for $g(Z)$. System paths not consonant with grammar and syntax are a set of measure zero having very many more members than the set of high probability paths.

A first, and central, assertion of this analysis is that the possibility of such differentiation—division into at least two equivalence classes—permits emergence from the prebiotic chemical soup of a living state engaging in cognitive behaviors. The differentiation of dynamic paths into high and low probability equivalence classes of behaviors represents the first groupoid structure, analogous to the 'underlying basic symmetry' of theoretical cosmology's 'big bang' formalism (CTC 2021). This inference, consonant with earlier studies regarding the origin of life, (e.g., Wallace and Wallace 2008; Wallace 2011a,b), elevates the asymptotic limit theorems of information theory to a central position in prebiotic studies.

Indeed, Feynman (2000) makes the direct argument that information itself is to be viewed as a form of free energy, using Bennett's clever ideal machine that turns a message directly into work. Here, however, we are concerned with an iterated, rather than a direct, construction.

F, taken as a free energy, then becomes subject to symmetry-breaking transitions as $g(Z)$ varies (Pettini 2007). These symmetry changes, however, are not as associated with physical phase transitions as represented by standard group algebras. Such symmetry changes represent transitions from playing one 'game' to playing another. For example, an organism may engage in foraging behaviors that trigger a predatory attack by another organism. Then the game changes from 'foraging' to 'escape'.

More generally, Stewart (2017) puts the matter as follows:

> Spontaneous symmetry-breaking is a common mechanism for pattern formation in many areas of science. It occurs in a symmetric dynamical system when a solution of the equations has a smaller symmetry group than the equations themselves... This typically happens when a fully symmetric solution becomes unstable and branches with less symmetry bifurcate.

Thus 'cognitive phase changes', as we characterize them here, involve shifts between equivalence classes of high probability developmental/behavioral pathways that are represented as groupoids. To reiterate, this represents a generalization of the group concept such that a product is not necessarily defined for every possible element pair, although multiple products with multiple identity elements are defined (Brown 1992; Cayron 2006; Weinstein 1996). Again, see the Mathematical Appendix for an outline of the theory.

Dynamic equations follow from invoking a first-order Onsager approximation akin to that of nonequilibrium thermodynamics (de Groot and Mazur 1984) in the gradient of an entropy measure constructed from the 'iterated free energy' F of Eq. (7.4). Recall that the central matter of the Onsager approximation is that $\partial Z/\partial t \approx \partial S/\partial Z$. The full algebraic thicket is

$$S(Z) \equiv -F(Z) + Z dF(Z)/dZ$$

$$\partial Z/\partial t \approx dS/dZ = f(Z)$$

$$f(Z) = Z d^2 F/dZ^2$$

$$g(Z) =$$

$$\frac{-C_1 Z - \left(\int \frac{f(Z)}{Z}dZ\right) Z + C_2 + \int f(Z)\,dZ}{RootOf\left(e^Q - h\left(-\frac{C_1 Z + \left(\int \frac{f(Z)}{Z}dZ\right)Z - C_2 - \left(\int f(Z)dZ\right)}{Q}\right)\right)} \quad (7.5)$$

where the last relation follows from an expansion of the third part of Eq. (7.5) using the second expression of Eq. (7.4).

Several important—and somewhat subtle—points follow from the last expressions in Eqs. (7.4) and (7.5).

1. 'It is easily seen that' the 'RootOf' constructions generalize the Lambert W-function (e.g., Yi et al. 2010; Mezo and Keady 2015).

2. Further, since 'RootOf' may have complex number solutions, the temperature analog $g(Z)$ enters the realm of the 'Fisher Zeros' characterizing phase transition in physical systems (e.g., Dolan et al. 2001; Fisher 1965; Ruelle 1964 Sec. 5).
3. Information sources are not microreversible, that is, palindromes are highly improbable, e.g., ' ot ' has far lower probability than ' to ' in English, so that there are no 'Onsager Reciprocal Relations' in higher dimensional systems. The necessity of groupoid symmetries appears to be driven by this directed homotopy.
4. There will typically be some delay in the rate of provision of Z, so that, in Eq. (7.5), for example, $f(Z) = \beta - \alpha Z(t)$—an exponential model having $Z(t) = (\beta/\alpha)(1 - \exp[-\alpha t])$—where $Z \to \beta/\alpha$ at a rate determined by α. Other dynamics are possible, such as the 'Arrhenius', $Z(t) = \beta \exp[-\alpha/t]$, with $f(Z) = (Z/\alpha)(\log(Z/\beta))^2$. However Z is constructed from the components $\mathscr{C}$, $\mathscr{D}$ and $\mathscr{M}$, so here, it is the scalar resource rate Z itself that counts.

Suppose, in the first expression of Eq. (7.5), it is possible to approximate the sum across the high probability paths with an integral, so that

$$\exp[-F/g(Z)] \approx \int_0^\infty \exp[-H/g(Z)]dH = g(Z) \qquad (7.6)$$

$g(Z)$ must be real-valued and positive. Then

$$F(Z) = -\log[g(Z)]g(Z)$$
$$g(Z) = -F(Z)/W_L[n, -F(Z)] \qquad (7.7)$$

where W_L is the 'simple' Lambert W-function that satisfies $W_L[n, x]$ $\exp[W_L[n, x]] = x$. It is real-valued only for $n = 0, -1$ and only over limited ranges of x in each case.

In theory, specification of any two of the functions f, g, h permits calculation of the third. h, however, is determined—fixed—by the internal structure of the larger system. Similarly, 'boundary conditions' C_1, C_2 are externally-imposed, further sculpting dynamic properties of the 'temperature' $g(Z)$, and f determines the rate at which the composite essential resource Z can be delivered. Both information and metabolic free energy resources are rate-limited.

7.5 Cognitive 'Reaction Rate'

For physical systems there will necessarily be a minimum temperature for punctuated activation of the particular set of dynamics associated with a given group structure. For cognitive processes, following the arguments of Eqs. (7.4) and (7.5), there will be a minimum necessary value of $g(Z)$ for onset of the next in a series of transitions. That is, at some $T_0 \equiv g(Z_0)$, having a corresponding information source uncertainty H_0, other groupoid phase transitions become manifest.

Taking a reaction rate perspective from chemical kinetics (Laidler 1987), we can write a rate expression as

$$L(Z) = \frac{\sum_{H_j > H_0} \exp[-H_j/g(Z)]}{\sum_k \exp[-H_k/g(Z)]} \tag{7.8}$$

If the sums can be approximated as integrals, then the system's rate of 'reaction' at resource rate Z can be written as

$$L(Z) \approx \frac{\int_{H_0}^{\infty} \exp[-H/g(Z)]dH}{\int_0^{\infty} \exp[-H/g(Z)]dH} = \exp[-H_0/g(Z)]$$

$$= \exp[H_0 W_L(n, -F)/F] \tag{7.9}$$

where $W_L(n, -F)$ is the Lambert W-function of order n in the free energy index $F = -\log[g(Z)]g(Z)$, and we enter a brave new world.

Figure 7.1 shows $L(F)$ vs. F, using Lambert W-functions of orders 0 and -1, respectively real-valued only on the intervals $x > -\exp[-1]$ and $-\exp[-1] < x < 0$.

The Lambert W-function is only real-valued for orders 0 and -1, and only for $F \leq \exp[-1]$. However, if $F > \exp[-1]$, then a bifurcation instability emerges, with a transition to complex-valued oscillations in cognition rate at higher values.

This development recovers what is essentially an analog to the Data Rate Theorem from control theory (Nair et al. 2007), in the sense that the requirement

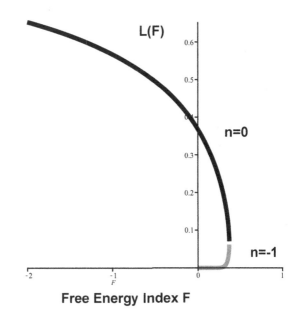

Fig. 7.1 Rate of cognition from Eq. (7.9) as a function of the iterated free energy measure F, taking $H_0 = 1$. The Lambert W-function is only real-valued for orders 0 and -1, and only if $F \leq \exp[-1]$. However, if $F > \exp[-1]$, then a bifurcation instability emerges, with a transition to complex-valued oscillations in cognition rate at higher values. Since F is driven by Z, there is a minimum resource rate for stability

$H > H_0$ in Eqs. (7.8) and (7.9) imposes stability constraints on F, the free energy analog, and by inference, on the resource rate index Z driving it.

7.6 Phase Transitions II: Signal Transduction and Noise

We can carry the basic results somewhat further if we again approximate the sum in Eq. (7.4) by an integral, that is taking $h(g(Z)) \approx g(Z)$, and setting $dZ/dt = f(Z) = \beta - \alpha Z(t)$. Then

$$g(Z) = -\frac{2\ln(Z)\, Z\beta - Z^2\alpha + 2C_1 Z - 2Z\beta + 2C_2}{2W_L\left(n,\, -\ln(Z)\, Z\beta + \frac{Z^2\alpha}{2} - C_1 Z + Z\beta - C_2\right)} \tag{7.10}$$

with, again, $L = \exp[-H_0/g(Z)]$, depending on the rate parameters α and β, the boundary conditions C_i, and the degree of the Lambert W-function. Proper choice of boundary conditions generates a classic signal transduction analog (e.g., Wallace 2020a, 2021c; Diamond et al. 2007). That is, since $Z \rightarrow \beta/\alpha$, we look at the cognition rate for fixed α and boundary conditions C_j as β increases. The result is shown in Fig. 7.2, for appropriate boundary conditions, taking the 'activation energies' H_0 at different values. Note the markedly higher cognition rate at the lower activation energy for a given 'temperature' $g(Z)$, a classic outcome relevant to the observations of Peeples and Rosen (2021) for biomolecular condensates.

Similar results follow if $\exp[-F/g(Z)] = h(g(Z)) \propto A_m g(Z)^m$, that is, if the function $h(g(Z))$ has a strongly dominant term of order $m > 0$.

Fig. 7.2 Classic signal transduction for the cognition rate from Eq. (7.10), setting $n = 0$, $\alpha = 1$, $C_1 = -2$, $C_2 = 2$, $H_0 = 1$, 15. β is taken as the 'arousal' measure. Boundary conditions are appropriate to a signal transduction model. Note, however, the singular dependence on 'activation energy' H_0, relevant to the observations of Peeples and Rosen (2021) for biomolecular condensates

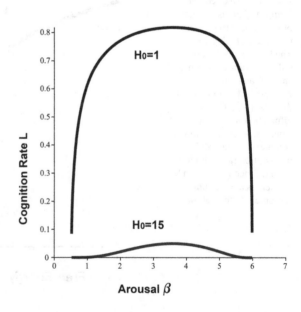

It is possible to infer the effects of 'noise' on the rate of cognition/reaction in this system by expanding the second expression in Eq. (7.5) as a stochastic differential equation (Protter 2005)

$$dZ_t = f(Z_t)dt + \sigma Z_t dB_t$$
$$= (\beta - \alpha Z_t)dt + \sigma Z_t dB_t \tag{7.11}$$

and then applying the Ito Chain Rule (Protter 2005) via Eq. (7.9) to derive the nonequilibrium steady state (nss) condition $< dL_t >= 0$, where the brackets indicate a time average. Here, dB_t represents ordinary Brownian noise and the second term is the standard expression for 'volatility' in the sense of financial engineering, with σ indexing the 'noise' strength.

The system is taken at the peak of Fig. 7.2 for $H_0 = 1$, with $\beta = 7/2$ and the other parameters the same as in that figure, leading to Fig. 7.3.

Two points. First, the nss relation $< dL_t >= 0$ generates bifurcation instabilities driven by increasing σ, in essence noise-driven phase transitions in the sense of Tian et al. (2017), Horsthemeke and Lefever (2006) and Van den Broeck et al. (1994, 1997).

Second, the two disconnected solution set graphs $\{Z, \sigma\}$ form disjoint equivalence classes that can be seen as constituting the foundation of an associated groupoid, whose symmetry-breakings represent system phase transitions.

Fig. 7.3 Fixing the system at the peak of figure 7.2 for $H_0 = 1$, so that $\beta = 7/2$ with other parameters the same, the imposition of 'noise' of strength σ via Eq. (7.11) on the nonequilibrium steady state cognition, so that $< dL_t >= 0$, produces noise-driven bifurcation instability phase transitions, based on the groupoid equivalence classes defined by the two disjoint solution sets

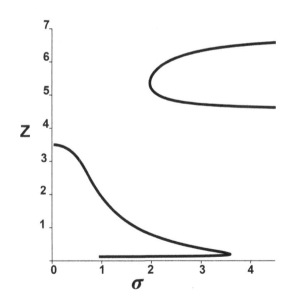

7.7 Discussion

The centrality of Wallace (2022a)—presented here as Chap. 1—was a three-fold classification of the major transitions in evolutionary biology. These are seen as collapsed into a relatively simple sequential progression in complexity:

1. Emergence of high vs low probability reaction/developmental paths.
2. System division into interior and exterior by various mechanisms.
3. Emergence of multiple, interacting tunable workspaces.

From that perspective, biomolecular condensates within cells might well represent 'fossil survivals' of a second stage in evolutionary process that was a precursor of membrane-separated components (Martin and Douliez 2021, and references therein). That is, within some highly condensed prebiotic soup, such condensates could assemble, disperse, and reassemble as-needed under prevailing selection pressures, perhaps based on heritable primitive 'scaffolding'.

That is, just as mitochondria and other membrane-bound organelles represent such 'fossils' within eukaryotic cells, biomolecular condensates may represent continuation of a much earlier generation of prebiotic systems, antedating even the presumed RNA World. Evolution, after all, tends to conserve what has worked in what emerges under subsequent selection.

We can, in fact, use this perspective to reconsider matters of reaction rate as follows.

Biomolecular condensates, as we characterize them here, are to be taken as 'remnant organisms' within eukaryotic cells, and thus must respond to 'selection pressure' signals from that embedding environment, the cell itself. Their singular nature is that, unlike under 'ordinary' selection pressures, extinction is not forever. More specifically, we return to the formalism of Sect. 2.5, assuming a possible 'spectrum' of biomolecular condensates within the cell, each operating with some cognition/reaction rate L_i, under constraints of 'resource rate X_i and time T_i such that

$$\partial X_i / \partial t = f_i(X_i)$$

$$\sum_i X_i = X$$

$$\sum_i T_i = T \tag{7.12}$$

We then impose a Lagrangian optimization on the spectrum via the rates L_i—a skeletal version of the ubiquitous 'fitness landscape'—as

$$\mathscr{L} = \sum_i L_i + \lambda \left(X - \sum_i X_i \right) + \mu \left(T - \sum_i T_i \right)$$

$$\partial \mathscr{L}/\partial X = \lambda$$

$$\partial L_i/\partial X_i = \lambda$$

$$\partial L_i/\partial T_i = \mu \qquad (7.13)$$

As in Sect. 2.5 we assume that $\partial X_i/\partial T_i = f_i(X_i(T_i)) \to 0$ as T_i increases. Some simple algebra finds

$$f_i(X_i) = \frac{\mu}{\lambda}$$

$$X_i = f_i^{-1}\left(\frac{\mu}{\lambda}\right) \qquad (7.14)$$

In consequence, a monotonic function $f_i(X_i)$ will generate a monotonic dependence on X_i on the shadow price ratio μ/λ that represents 'selection pressure' demands imposed by the cell-as-environment. Again, assuming $f_i(X_i) = \beta_i - \alpha_i X_i$, the exponential model, gives

$$X_i = \frac{\beta_i - \mu/\lambda}{\alpha_i} \qquad (7.15)$$

Thus 'environmental signals' from the embedding cell act as shadow prices to determine the rates at which resources are to be delivered across the spectrum of possible biomolecular condensates. Indeed, the parameters β_i, α_i are also likely tunable according to cellular state.

Again, as above, the optimization treatment of Eq. (7.12) et seq. can be seen as condensation of the more familiar 'fitness landscape' arguments of Gavrilets (2010) and associated references.

The full range of our formal developments can be seen as translating the language of 'chemical reaction rate' into the information-theory dialect of 'cognition rate' in these pre-membrane systems. In particular, Sects 7.4 and 7.6 suggest searching for groupoid symmetry-breaking and other phase transition dynamics in the assembly, reassembly, and dynamic function of currently-observed biomolecular condensates, if only for possible insights into, and reconstructions of, no longer observable prebiotic structures.

7.8 Mathematical Appendix: Groupoids

Following Brown (1992) closely, consider a directed line segment in one component, written as the source on the left and the target on the right.

$$\bullet \longrightarrow \bullet$$

Two such arrows can be composed to give a product **ab** if and only if the target of **a** is the same as the source of **b**

Brown puts it this way,

One imposes the geometrically obvious notions of associativity, left and right identities, and inverses. Thus a groupoid is often thought of as a group with many identities, and the reason why this is possible is that the product **ab** is not always defined.

We now know that this apparently anodyne relaxation of the rules has profound consequences... [since] the algebraic structure of product is here linked to a geometric structure, namely that of arrows with source and target, which mathematicians call a directed graph.

Cayron (2006) elaborates this as follows,

A group defines a structure of actions without explicitly presenting the objects on which these actions are applied. Indeed, the actions of the group G applied to the identity element e implicitly define the objects of the set G by ge = g; in other terms, in a group, actions and objects are two isomorphic entities. A groupoid enlarges the notion of group by explicitly introducing, in addition to the actions, the objects on which the actions are applied. By this approach, many identities may exist (they correspond to the actions that leave an object invariant).

It is of particular importance that equivalence class decompositions permit construction of groupoids in a highly natural manner.

Weinstein (1996) and Golubitsky and Stewart (2006) provide more details on groupoids and on the relation between groupoids and bifurcations.

An essential point is that, since there are no necessary products between groupoid elements, 'orbits', in the usual sense, disjointly partition groupoids into 'transitive' subcomponents.

References

Atlan H., and I. Cohen. 1998. Immune information, self-organization, and meaning. International Immunology 10:711–717.

Brown, R. 1992. Out of line. Royal Institute Proceedings 64:207–243.

Brown, R., P. Higgins, and R. Sivera. 2011. *Nonabelian Algebraic Topology: Filtered Spaces, Crossed Complexes, Cubical Homotopy Groupoids*. EMS tracts in mathematics, vol. 15.

Cayron, C. 2006. Groupoid of orientational variants. Acta Crystalographica Section A A62:21040.

Champagnat, N., R. Ferriere, and S. Meleard. 2006. Unifying evolutionary dynamics: from individual stochastic process to macroscopic models. Theoretical Population Biology 69:297–321.

Cover, T., and J. Thomas. 2006. *Elements of Information Theory*, 2nd ed. New York: Wiley.

CTC. 2021. http://www.ctc.cam.ac.uk/outreach/origins/cosmic_structures_one.php

de Groot, S., and P. Mazur. 1984. *Nonequilibrium Thermodynamics*. New York: Dover.

Dembo, A., and O. Zeitouni, 1998. *Large Deviations and Applications*, 2nd ed. New York: Springer.

Diamond, D., A. Campbell, C. Park, J. Halonen, and P. Zoladz. 2007. The temporal dynamics model of emotional memory processing. Neural Plasticity 2007:60803. https://doi.org/10.1155/2007/60803

Dolan, B., W. Janke, D. Johnston, and M. Stathakopoulos. 2001. Thin Fisher zeros. Journal of Physics A 34:6211–6223.

Feynman, R. 2000. *Lectures in Computation*. Boulder: Westview Press.

Fisher, M. 1965. *Lectures in Theoretical Physics*, Vol. 7. Boulder: University of Colorado Press.

Gavrilets, S. 2010. High-dimensional fitness landscapes and speciation. In *Evolution: The Extended Synthesis*, ed. Massimo Pigliucci and Gerd B. Müller. Cambridge: MIT Press.

Golubitsky, M., and I. Stewart. 2006. Nonlinear dynamics and networks: the groupoid formalism. Bulletin of the American Mathematical Society 43:305–364.

Hatcher, A. 2001. *Algebraic Topology*. New York: Cambridge University Press.

Horsthemeke, W., and R. Lefever. 2006. *Noise-Induced Transitions*. Theory and Applications in Physics, Chemistry, and Biology, Vol. 15. New York: Springer.

Hoyrup, M. 2013. Computability of the ergodic decomposition. Annals of Pure and Applied Logic 164:542–549.

Jolliffe, I. 2002. *Principal Component Analysis*. New York: Springer.

Khinchin, A. 1957. *Mathematical Foundations of Information Theory*. New York: Dover.

Laidler, K. 1987. *Chemical Kinetics*, 3rd ed. New York: Harper and Row.

Landau, L., and E. Lifshitz. 2007. *Statistical Physics*, 3rd ed., Part 1. New York: Elsevier.

Lin Y., J. Forman-Kay, and H. Chan. 2018. Theories for sequence-dependent phase behaviors of biomolecular condensates. Biochemistry 57:2499–2508. https://doi.org/10.1021/acs.biochem.8b00058

Lyon, A., W. Peeples, and M. Rosen. 2021. A framework for understanding functions of biomolecular condensates on molecular cellular scales. Nature Reviews Molecular Cell Biology 22:215–235.

Marshall, J. 2014. The genetic code. PNAS 111:5760.

Martin, E., et al. 2020. Valence and patterning of aromatic residues determine the phase behavior of prion-prionlike domains. Science 367:694–699. https://doi.org/10.1126/science.aaw8653

Martin, N., and J. Douliez. 2021. Fatty acid vesicles and coacervates as model prebiotic protocells. ChemSystemsChem 3:e2100024.

Maturana, H., and F. Varela. 1980. *Autopoiesis and Cognition: The Realization of the Living*. Boston: Reidel.

Mezo I., and G. Keady. 2015. Some physical applications of generalized Lambert functions. arXiv:1505.01555v2 [math.CA] 22 Jun 2015.

Nair, G., F. Fagnani, S. Zampieri, and R. Evans. 2007. Feedback control under data rate constraints: an overview. Proceedings of the IEEE 95:108137.

Peeples, W., and M. Rosen. 2021. Mechanistic dissection of increased enzymatic rate in a phase separated component. Nature Chemical Biology 17:693–702.

Pettini, M. 2007. *Geometry and Topology in Hamiltonian Dynamics and Statistical Mechanics*. New York: Springer.

Protter, P. 2005. *Stochastic Integration and Differential Equations: A New Approach*, 2nd ed. New York: Springer.

Ruelle, D. 1964. Cluster property of the correlation functions of classical gases. Reviews of Modern Physics 38:580–584.

Sang, D., T. Shu, C. Pantoja, A. Ibanez de Opakua, M. Zweckstetter, and L. Holt. 2022. Condensed-phase signaling can expand kinase specificity and respond to macromolecular crowding. Molecular Cell 82:P3693-3711.E10. https://doi.org/10.1016/j.molcel.2022.08.016

Stewart, I. 2017. Spontaneous symmetry-breaking in a network model for quadruped locomotion. International Journal of Bifurcation and Chaos 14:1730049.

Tian, C., L. Lin, and L. Zhang. 2017. Additive noise driven phase transitions in a predator-prey system. Applied Mathematical Modelling 46:423–432.

Van den Broeck, C., J. Parrondo, and R. Toral. 1994. Noise-induced nonequilibrium phase transition. Physical Review Letters 73:3395–3398.

Van den Broeck, C., J. Parrondo, R. Toral, and R. Kawai. 1997. Nonequilibrium phase transitions induced by multiplicative noise. Physical Review E 55:4084–4094.

Vernon, R., et al. 2018. Pi-Pi contacts are an overlooked protein feature relevant to phase separation. eLife 7:e31486. https://doi.org/10.7554/eLife.31486

Wallace, R. 2005. *Consciousness: A Mathematical Treatment of the Global Neuronal Workspace Model*. New York: Springer.

Wallace, R. 2011a. On the evolution of homochirality. Comptes Rendus Biologies 334:263–268.

Wallace, R. 2011b. Structure and dynamics of the 'protein folding code' inferred using Tlusty's topological rate distortion approach. BioSystems 103:18–26.

Wallace, R. 2012. Consciousness, crosstalk, and the mereological fallacy: an evolutionary perspective. Physics of Life Reviews 9:426–453.

Wallace, R. 2014. A new formal perspective on 'Cambrian Explosions'. Comptes Rendus Biologies 337:1–5.

Wallace, R. 2015. *An Ecosystem Approach to Economic Stabilization: Escaping the Neoliberal Wilderness*. New York: Routledge.

Wallace, R. 2017. *Computational Psychiatry: A Systems Biology Approach to the Epigenetics of Mental Disorders*. New York: Springer.

Wallace, R. 2018. New statistical models of nonergodic cognitive systems and their pathologies. Journal of Theoretical Biology 436:72–78.

Wallace, R. 2020a. On the variety of cognitive temperatures and their symmetry-breaking dynamics. Acta Biotheoretica 68:421–439. https://doi.org/10.1007/s10441-019-09375-7

Wallace, R. 2020b. *Cognitive Dynamics on Clausewitz Landscapes: The Control and Directed Evolution of Organized Conflict*. New York: Springer.

Wallace, R. 2020c. Signal transduction in cognitive systems: origin and dynamics of the inverted-U/U dose-response relation. Journal of Theoretical Biology 504:110377.

Wallace, R. 2021a. How AI founders on adversarial landscapes of fog and friction. Journal of Defense Modeling and Simulation 19. https://doi.org/10.1177/1548512920962227

Wallace, R. 2021b. Toward a formal theory of embodied cognition. BioSystems 202:104356.

Wallace, R. 2021c. Embodied cognition and its pathologies: the dynamics of institutional failure on wickedly hard problems. Communications in Nonlinear Science and Numerical Simulation 95:105616.

Wallace, R. 2022. Major transitions as groupoid symmetry-breaking in nonergodic prebiotic, biological and social information systems. Acta Biotheoretica 70:27. https://doi.org/10.1007/s10441-022-09451-5

Wang, J., et al. 2018. A molecular grammar governing the driving forces for phase separation of prion-like RNA binding proteins. Cell 174:688–699.e616. https://doi.org/10.1016/j.cell.2018.06.006

Weinstein, A. 1996. Groupoids: unifying internal and external symmetry. Notices of the American Mathematical Association 43:744–752.

Yi, S., P.W. Nelson, and A.G. Ulsoy. 2010. *Time-Delay Systems: Analysis and Control Using the Lambert W Function*. New Jersey: World Scientific.

Chapter 8
Evolutionary Exaptation: Shared Interbrain Activity in Social Communication

People are embedded in social interaction that shapes their brains throughout lifetime. Instead of emerging from lower-level cognitive functions, social interaction could be the default mode via which humans communicate with their environment.
— Hari et al. (2015)

The challenge for the study of brain-to-brain coupling is to develop detailed models of the dynamical interaction that can be applied at the behavioural levels and at the neural levels.
— Hasson and Frith (2016)

...[A] deeper understanding of inter-brain dynamics may provide unique insight into the neural basis of collective behavior that gives rise to a broad range of economic, political, and sociocultural activities that shape society.
— Kingsbury and Hong (2020)

8.1 Introduction

A recent elegant study of a bat population by Rose et al. (2021) finds that bidirectional interbrain activity patterns are a feature of their socially interactive behaviors, and that such shared interbrain activity patterns likely play an important role in social communication between group members. Sliwa (2021) summarizes that work, and parallel material by Baez-Mendoza et al. (2021) on macaques. Kingsbury et al. (2019), in a particularly deep analysis, studied correlations in brain activity between socially interacting mice, finding strong structuring by dominance relations.

Rose et al. are careful to cite the large and growing human literature on brain-to-brain coupling in social interaction, including Hasson et al. (2012), Barraza et al. (2020), Kuhlen et al. (2017), Perez et al. (2017), Stolk et al. (2014), and Dikker et al. (2017). As Rose et al. put it, a wide range of species naturally interact in groups and exhibit a diversity of social structures and forms of communication involving similarities and differences in neural repertoires for social communication.

On far longer time scales, Abraham et al. (2020) found that concordance in parent and offspring cortico-basal ganglia white matter connectivity varies by parental history of major depressive disorder and early parental care. As they put it,

Social behavior is transmitted cross-generationally through coordinated behavior within attachment bonds. Parental depression and poor parental care are major risks for disruptions of such coordination and are associated with offspring's psychopathology and interpersonal dysfunction... [Study] showed diminished neural concordance among dyads with a depressed parent and that better parental care predicted greater concordance, which also provided a protective buffer against attenuated concordance among dyads with a depressed parent... [Such] disruption may be a risk factor for intergenerational transmission of psychopathology. Findings emphasize the long-term role of early caregiving in shaping neural concordance among at-risk and affected dyads.

Indeed, a broad spectrum of Holocaust studies (Dashorst et al. 2019) has followed intergenerational transmission of psycho- and other pathologies, including, but not limited to, patterns of brain function.

Here, following Wallace (2022a,b), we examine 'shared interbrain activity patterns' from the perspectives of recent developments in control and information theories, using the asymptotic limit theorems of those disciplines to develop probability models that might be converted to statistical tools of value in future observational and empirical studies of the phenomena at various time scales. There is, after all, a very long tradition of using control theory ideas in psychological research. See the review by Henry et al. (2022) for a deep and cogent summary.

Shared interbrain activity patterns are concrete representations—indeed, instantiations—of information transmission within a group, and, as Dretske (1994) indicates, the properties of any transfer of information are strongly constrained by the asymptotic limit theorems of information theory, in the same sense that the Central Limit Theorem imposes constraints leading to useful statistical models of supposedly 'random' phenomena.

We begin with 'simple' correlation of brain activity between individuals in social interaction, and then move on to more complex models of joint cognition across individuals and/or 'workgroups', in a large sense.

8.2 Correlation

The elegant paper by Kingsbury et al. (2019) explores correlated neural activity and the encoding of behavior across the brains of socially-interacting mice. Two central findings of that work are shown in Fig. 8.1, (as adapted from their Figs. 8.2 and 8.8). The top row of Fig. 8.1 shows time series of brain activity in two mice, first with, and then without, direct contact. Correlation between the signals is much higher during social interaction. The lower part indicates that correlations rise with difference in status between animals.

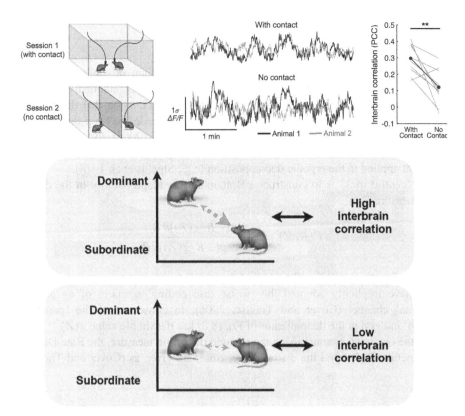

Fig. 8.1 Adapted from Kingsbury et al. (2019), figures 1 and 8. Top: correlation between brain activity is much higher during social interaction. Bottom: Correlation is much higher between discordantly dominant animals

Here, we derive these results using perhaps the simplest possible dynamic model, one based on a principal asymptotic limit theorem of information theory, the Rate Distortion Theorem.

We adapt the introductory model of Wallace (2020b, Sec. 1.3), focused on interacting institutions under conditions of conflict analogous to status disjunction.

Suppose we have developed—or been given—a robust scalar measure of social dominance between pairs of individuals, Z. How does Z affect 'correlation', in a large sense, during interactions?

A 'dominant' partner transmits signals to a 'subordinate' in the presence of noise, sending a sequence of signals $U_i = \{u^i_1, u^i_2 \ldots\}$ that,—again, in the presence of noise—is received as $\hat{U}^i = \{\hat{u}^i_1, \hat{u}^i_2, \ldots\}$. We take the U^i as sent with probabilities $P(U^i)$, and define a scalar distortion measure between U^i and $\hat{U}^i$ as $d(U^i, \hat{U}^i)$, defining an average distortion D as

$$D \equiv \sum_i P(U^i) d(U^i, \hat{U}^i) \tag{8.1}$$

Following Wallace (2020b, Sec. 1.3) closely, it is then possible to apply a rate distortion argument. The Rate Distortion Theorem—for stationary, ergodic systems—states that there is a convex Rate Distortion Function that determines the minimum channel capacity, written $R(D)$, that is needed to keep the average distortion below the limit D. See Cover and Thomas (2006) for details. The theory can be extended, with some difficulty, to nonergodic sources via an infimum argument applied to the ergodic decomposition (e.g., Shields et al. 1978).

The 'central trick' is to construct a Boltzmann pseudoprobability in the dominance measure Z as

$$dP(R, Z) = \frac{\exp[-R/g(Z)] dR}{\int_0^\infty \exp[-R/g(Z)] dR} \tag{8.2}$$

where the function $g(Z)$ is unknown and must be determined from first principles.

We have implicitly adopted the 'worst case coding' scenario of an analog 'Gaussian' channel (Cover and Thomas 2006). In consequence, the 'partition function' integral in the denominator of Eq. (8.2) has the simple value $g(Z)$.

For the Gaussian channel using the squared distortion measure, the Rate Distortion Function $R(D)$, and the distortion measure, are given as (Cover and Thomas 2006)

$$R(D) = \frac{1}{2} \log_2[\sigma^2/D]$$

$$D = \sigma^2 2^{-2R} \tag{8.3}$$

If $D \geq \sigma^2$, then $R = 0$.

From these relations, using Eq. (8.2), and after some manipulation, it is possible to calculate the ' average distortion' $< D >$ as

$$< D >= \frac{\sigma^2}{\log(4) g(Z) + 1} \tag{8.4}$$

For the 'natural channel'

$$D = \frac{\sigma^2}{1 + R}$$

$$< D >= \frac{\sigma^2 e^{\frac{1}{g(Z)}} \mathrm{Ei}_1\left(\frac{1}{g(Z)}\right)}{g(Z)} \tag{8.5}$$

where Ei_1 is the exponential integral of order 1.

What, then, is $g(Z)$?

Here we abduct more formalism from statistical mechanics, using the 'partition function' in Eq. (8.2) to define an 'iterated free energy' as

$$\exp[-F/g(Z)] = \int_0^\infty \exp[-R/g(Z)]dR = g(Z)$$

$$F = -\log[g(Z)]g(Z)$$

$$g(Z) = \frac{-F(Z)}{W(n, -F(Z))} \qquad (8.6)$$

where $W(n, x)$ is the Lambert W-function of order n solving the relation $W(n, x)\exp[W(n, x)] = x$ and is real-valued only for $n = 0, -1$ over limited ranges. These are, respectively, for $n = 0, x > -\exp[-1]$, and for $n = -1, -\exp[-1] < x < 0$. These conditions are important and impose themselves on the expressions for $< D >$.

The next step is to define an 'iterated entropy' in the standard manner as the Legendre transform of the 'free energy' analog F, and from it, impose a simple first-order version of the Onsager treatment of nonequilibrium thermodynamics (de Groot and Mazur 1984), in the context of a nonequilibrium steady state, so that $dZ/dt = 0$. The relations are then

$$S(Z) \equiv -F(Z) + ZdF(Z)/dZ$$

$$dZ/dt \propto dS/dZ = Zd^2F/dZ^2 = 0$$

$$F(Z) = C_1 Z + C_2 \qquad (8.7)$$

where the C_i are appropriate boundary conditions and $g(Z)$ is then given by the last expression in Eq. (8.6).

For the Gaussian channel, taking the Lambert W-function of order zero, and setting $C_1 = -1$, $C_2 = 1/10$, gives Fig. 8.2.

For the 'natural' channel, again with the W-function of order zero, taking the same values for the C_i gives Fig. 8.3.

Higher status disjunction, in this model, implies closer coupling between individuals.

Similar results will follow for all possible Rate Distortion Functions that must be both convex in D and zero-valued for $D \geq \sigma^2$ (Cover and Thomas 2006).

A fairly elementary model of social interaction under dominance relations, based on a somewhat counter-intuitive dynamic adaptation of the Rate Distortion Theorem, produces results consistent with the empirical observations of Kingsbury et al. (2019). Extension of the model to both nonergodic and nonstationary conditions more likely to mirror real-world conditions, or at least going beyond the nonequilibrium steady state assumption, requires further work.

The model can, in theory, be extended using the ergodic decomposition of a nonergodic process using the methods of Shields et al. (1978).

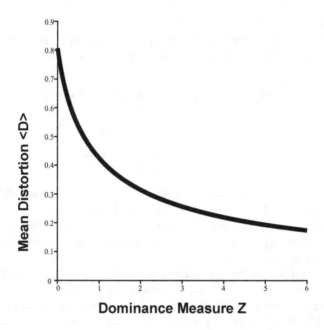

Fig. 8.2 Mean distortion $< D >$ for the Gaussian channel vs. the dominance index Z. Higher status disjunction implies closer coupling during social interaction, according to this model

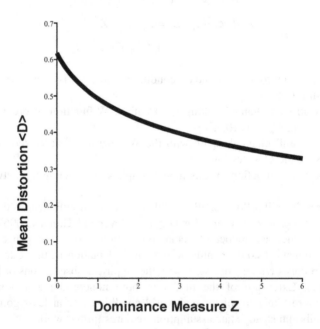

Fig. 8.3 Mean distortion $< D >$ for the 'natural' channel vs. the dominance index Z. Again, higher status disjunction implies closer coupling during social interaction

Otherwise, we are constrained to adiabatically, piecewise, stationary ergodic (APSE) systems that remain as close to ergodic and stationary as needed for the Rate Distortion Theorem to work, much like the Born-Oppenheimer approximation of molecular physics in which rapid electron dynamics are assumed to quasi-equilibrate about much slower nuclear oscillations, allowing calculations using 'simple' quantum mechanics models.

8.3 Cognition

Cognition is not correlation, and requires more general address. Indeed, cognition has become a kind of shibboleth in theoretical biology, seen by some as the fundamental characterization of the living state at and across its essential scales and levels of organization (Maturana and Varela 1980). In this regard, a central inference by Atlan and Cohen (1998), in their study of the immune system, is that cognition, via mechanisms of choice, demands reduction in uncertainty, implying the existence of information sources 'dual' to any cognitive process. The argument is unambiguous and direct, and serves as the foundation to our general approach.

A first step is to view 'information' as a biological and social resource matching the importance of metabolic free energy and other overtly material agents and agencies.

Information and Other Resources

Here, we must move beyond 'simple' measures of dominance between individuals in social interaction.

At least three resource streams are required by any cognitive entity facing real-time, real-world challenges. The first is measured by the rate at which information can be transmitted between elements within the entity, determined as an information channel capacity, say $\mathscr{C}$ (Cover and Thomas 2006). The second resource stream is sensory information regarding the embedding environment—here, primarily social interaction—available at a rate $\mathscr{Q}$. It is along this channel that 'neural representations' will be shared.

The third regards material resources, including metabolic free energy—in a large sense—available at a rate $\mathscr{M}$.

These three rates may well—but not necessarily—interact, a matter characterized as a 3 by 3 matrix analogous to, but not the same as, a simple correlation matrix. Let us write this as $\mathbf{Z}$.

An n-dimensional square matrix has n scalar invariants r_i defined by the relation

$$p(\gamma) = \det[\mathbf{Z} - \gamma\mathbf{I}] =$$

$$\gamma^n - r_1\gamma^{n-1} + r_2\gamma^{n-2} - \ldots + (-1)^n r_n \qquad (8.8)$$

I is the n-dimensional identity matrix, det the determinant, and γ a real-valued parameter. The first invariant is usually taken as the matrix trace, and the last as $\pm$ the determinant.

These scalar invariants make it possible to project the full matrix down onto a single scalar index $Z = Z(r_1, \ldots, r_n)$ retaining much of the basic structure, analogous to conducting a principal component analysis. The simplest index might be $Z = \mathscr{C} \times \mathscr{Q} \times \mathscr{M}$—the matrix determinant for a system without crossinteraction. However, scalarization must be appropriate to the individual circumstance studied, and there will almost always be important cross-interactions between resource streams.

Clever scalarization, however, enables approximate reduction to a one dimensional system.

Taking $\mathscr{M}$ out of the equation—equalizing it across the social structure—might generate two independent 'orthogonal' indices, for example the determinant and the trace of the 'interaction matrix' separately, so that Z becomes a two dimensional vector. General expansion of Z into vector form leads to difficult multidimensional dynamic equations (e.g., Wallace 2021c). See the Chapter Mathematical Appendix for details.

Cognition and Information

Only in the case that cross-sectional and longitudinal means are the same can information source uncertainty be expressed as a conventional Shannon 'entropy' (Khinchin 1957; Cover and Thomas 2006). Here, we only require that source uncertainties converge for sufficiently long paths, not that they fit some functional form. It is the values of those uncertainties that will be of concern so that we study 'Adiabatically Piecewise Stationary' (APS) systems, in the sense of the Born-Oppenheimer approximation for molecular systems that assume nuclear motions are so slow in comparison with electron dynamics that they can be effectively separated, at least on appropriately chosen trajectory 'pieces' that may characterize the various phase transitions available to such systems. Extension of this work to nonstationary circumstances remains to be done.

This approximation can be carried out via a fairly standard Morse Function iteration (Pettini 2007).

The systems of interest here are composed of cognitive submodules that engage in crosstalk. At every scale and level of organization all such submodules are constrained by both their own internals and developmental paths and by the persistent regularities of the embedding environment, including the cognitive intent of colleagues, in a broad sense, and the regularities of 'grammar' and 'syntax' imposed by the embedding social structure.

Further, there are structured uncertainties imposed by the large deviations possible within that environment, again including the behaviors of adversaries who may be constrained by quite different developmental trajectories and 'punctuated equilibrium' evolutionary transitions.

Recapitulating somewhat the arguments of Wallace (2018, 2020a), the Morse Function construction assumes a number of interacting factors:

- As Atlan and Cohen (1998) argue, cognition requires choice that reduces uncertainty. Such reduction in uncertainty directly implies the existence of an information source 'dual' to that cognition at each scale and level of organization. The argument is unambiguous and sufficiently compelling.
- Cognitive physiological processes, like the immune and gene expression systems, are highly regulated, in the same sense that 'the stream of consciousness' flows between cultural and social 'riverbanks'. That is, a cognitive information source X_i is generally paired with a regulatory information source X^i.
- Environments (in a large sense), also have sequences of very high and very low probability: night follows day, hot seasons follow cold, and so on.
- 'Large deviations', following Champagnat et al. (2006) and Dembo and Zeitouni (1998), also involve sets of high probability developmental pathways, often governed by 'entropy' analog laws that imply the existence of another information source.

Full system dynamics must then be characterized by a joint, nonergodic information source uncertainty

$$H(\{X_i, X^i\}, V, L_D) \tag{8.9}$$

defined path-by-path and not represented as an 'entropy' function (Khinchin 1957). Consequently, each path will have it's own H-value, but the functional form of that value is not specified in terms of underlying probability distributions.

The set $\{X_i, X^i\}$ includes the internal interactive cognitive dual information sources of the system of interest and their associated regulators, V is taken as the information source of the embedding environment. This may include the actions and intents of adversaries/symbionts/colleagues, as well as 'weather'. L_D is the information source of the associated large deviations possible to the system.

Again, we are projecting the spectrum of essential resources onto a scalar rate Z.

The underlying equivalence classes of developmental or dynamic system paths used to define groupoid symmetries can be defined fully in terms of the magnitude of individual path source uncertainties, $H(x^j)$, $x^j \equiv \{x_0^j, x_1^j, \ldots x_n^j, \ldots\}$) alone.

Recall the central conundrum of the ergodic decomposition of nonergodic information sources. It is formally possible to express a nonergodic source as the composition of a sufficient number of ergodic sources, much as it is possible to reduce planetary orbits to a Fourier sum of circular epicycles, obscuring the basic dynamics. Hoyrup (2013) discusses the problem further, finding that ergodic decompositions are not necessarily computable. Here, we need focus only on the values of the source uncertainties associated with dynamic paths.

8.4 Dynamics

The next step is to build an iterated 'free energy' Morse Function (Pettini 2007) from a Boltzmann pseudoprobability, based on enumeration of high probability developmental pathways available to the system, taking $j = 1, 2, \ldots$, so that

$$P_j = \frac{\exp[-H_j/g(Z)]}{\sum_k \exp[-H_k/g(Z)]} \tag{8.10}$$

where H_j is the source uncertainty of the path j, which—again—we do not assume to be given as a 'Shannon entropy' since we are no longer restricted to ergodic sources.

The essential point is the ability to divide individual paths into two equivalence classes, a small set of high probability paths consonant with an underlying 'grammar' and 'syntax', and a much larger set of vanishingly low probability, a set of measure zero.

The temperature-analog characterizing the system, written as $g(Z)$ in Eq. (8.10), can be calculated—or at least approximated—via a first-order Onsager nonequilibrium thermodynamic approximation built from the partition function, i.e., the denominator of Eq. (8.10) (de Groot and Mazur 1984).

We define the 'iterated free energy' Morse Function as

$$\exp[-F/g(Z)] \equiv \sum_k \exp[-H_k/g(Z)] \equiv h(g(Z))$$

$$F(Z) = -\log[h(g(Z))]g(Z) \tag{8.11}$$

where the sum is over all possible high probability developmental paths of the system, again, those consistent with an underlying grammar and syntax. Again, system paths not consonant with grammar and syntax constitute a set of measure zero that is very much larger than the set of high probability paths.

Feynman (2000) makes the direct argument that information itself is to be viewed as a form of free energy, using Bennett's 'ideal machine' that turns a message into work. Here, we invoke an iterated—rather than a direct—free energy construction.

F, taken as a free energy, then becomes subject to symmetry-breaking transitions as $g(Z)$ varies (Pettini 2007). These symmetry changes, however, are not as associated with physical phase transitions as represented by standard group algebras. Cognitive phase changes involve shifts between equivalence classes of high probability developmental pathways to be represented as *groupoids*, a generalization of the group concept where a product is not necessarily defined for every possible element pair (Brown 1992; Cayron 2006; Weinstein 1996). See the Chapter Mathematical Appendix for an outline of the theory.

The disjunction described above—into high and low probability equivalence classes representing paths consonant with, or discordant from, underlying grammar and syntax—should be seen as the primary 'groupoid phase transition' affecting cognitive systems. It is essentially the biological 'big bang' of Maturana and Varela (1980). Think about this carefully.

Dynamic equations follow from invoking a first-order Onsager approximation akin to that of nonequilibrium thermodynamics (de Groot and Mazur 1984) in the gradient of an entropy measure constructed from the 'iterated free energy' F of Eq. (8.11):

$$S(Z) \equiv -F(Z) + Z dF(Z)/dZ$$

$$\partial Z/\partial t \approx dS/dZ = f(Z)$$

$$f(Z) = Z d^2 F/dZ^2$$

$$g(Z) =$$

$$\frac{-C_1 Z - \left(\int \frac{f(Z)}{Z} dZ\right) Z + C_2 + \int f(Z) dZ}{RootOf\left(e^Q - h\left(-\frac{C_1 Z + \left(\int \frac{f(Z)}{Z} dZ\right) Z - C_2 - \left(\int f(Z) dZ\right)}{Q}\right)\right)} \tag{8.12}$$

where the last relation follows from an expansion of the third part of Eq. (8.12) using the second expression of Eq. (8.11). C_1 and C_2 are two constants in the indefinite integral of the second derivative of $F(Z)$, and Q is the independent variable of the function being taken roots.

Three important—and somewhat subtle—points:

1. The 'RootOf' construction generalizes the Lambert W-function (e.g., Yi et al. 2010; Mezo and Keady 2015). This leads to deep waters: since 'RootOf' may have complex number solutions, the temperature analog $g(Z)$ enters the realm of the 'Fisher Zeros' characterizing phase transition in physical systems (e.g., Dolan et al. 2001; Fisher 1965; Ruelle 1964 Sec. 5).
2. Information sources are not microreversible, that is, palindromes are highly improbable, e.g., ' eht ' has far lower probability than ' the ' in English. In consequence, there are no 'Onsager Reciprocal Relations' in higher dimensional systems. The necessity of groupoid symmetries appears to be driven by this directed homotopy.
3. Typically, it is necessary to impose a delay in provision of Z, so that, for example, $dZ/dt = f(Z) = \beta - \alpha Z(t)$ and $Z \to \beta/\alpha$ at a rate determined by α.

Suppose, in the first of Eq. (8.11), it is possible to approximate the sum with an integral, so that

$$\exp[-F/g(Z)] \approx \int_0^\infty \exp[-H/g(Z)]dH = g(Z) \tag{8.13}$$

$g(Z)$ must be real-valued and positive. Then

$$F(Z) = -\log[g(Z)]g(Z)$$

$$g(Z) = -F(Z)/W_L[n, -F(Z)] \tag{8.14}$$

Again, W_L is the 'simplest' Lambert W-function that satisfies $W_L[n, x]$ $\exp[W_L[n, x]] = x$. It is real-valued only for $n = 0, -1$ and only over limited ranges of x in each case.

In theory, specification of any two of the functions f, g, h permits calculation of the third. h, however, is determined—fixed—by the internal structure of the larger system. Similarly, 'boundary conditions' C_1, C_2 are externally-imposed, further sculpting dynamic properties of the 'temperature' $g(Z)$, and f determines the rate at which the composite essential resource Z can be delivered. Both information and metabolic free energy resources are rate-limited.

8.5   Cognition Rate

For phase transitions in physical systems, there is generally a minimum temperature for punctuated activation of the dynamics associated with given group structure— underlying symmetry changes associated with the transitions of ice to water to steam, and so on. For cognitive processes, following the arguments of Eq. (8.5), there will be a minimum necessary value of $g(Z)$ for onset of the next in a series of transitions. That is, at some $T_0 \equiv g(Z_0)$, having a corresponding information source uncertainty H_0, a second groupoid phase transition becomes manifest.

Taking a reaction rate perspective from chemical kinetics (Laidler 1987), we can write an expression for the rate of cognition as

$$L(Z) = \frac{\sum_{H_j > H_0} \exp[-H_j/g(Z)]}{\sum_k \exp[-H_k/g(Z)]} \tag{8.15}$$

If the sums can be approximated as integrals, then the system's rate of cognition at resource rate index Z can be written as

$$L(Z) \approx \frac{\int_{H_0}^\infty \exp[-H/g(Z)]dH}{\int_0^\infty \exp[-H/g(Z)]dH} = \exp[-H_0/g(Z)]$$

$$= \exp[H_0 W_L(n, -F)/F] \tag{8.16}$$

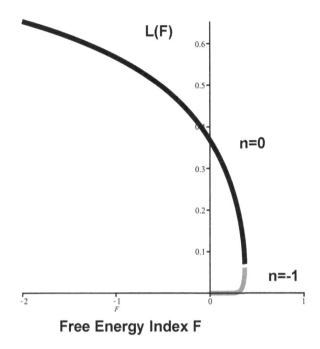

Free Energy Index F

Fig. 8.4 Rate of cognition from Eq. (8.16) as a function of the iterated free energy measure F, taking $H_0 = 1$. The Lambert W-function is only real-valued for orders 0 and -1, and only if $F < \exp[-1]$. However, if $0 < F < \exp[-1]$, then a bifurcation instability emerges, with a transition to complex-valued oscillations in cognition rate at higher values. Since F is driven by Z, there is a minimum resource rate for stability

where $W_L(n, -F)$ is the Lambert W-function of order n in the free energy index $F = -\log[g(Z)]g(Z)$.

Figure 8.4 shows $L(F) vs. F$, using Lambert W-functions of orders 0 and -1, respectively real-valued only on the intervals $x > -\exp[-1]$ and $-\exp[-1] < x < 0$.

The Lambert W-function is only real-valued for orders 0 and -1, and only for $F < \exp[-1]$. However, if $0 < F < \exp[-1]$, then a bifurcation instability emerges, with a transition to complex-valued oscillations in cognition rate at higher values.

This development recovers what is essentially an analog to the Data Rate Theorem from control theory (Nair et al. 2007 and the Chapter Mathematical Appendix), in the sense that the requirement $H > H_0$ in Eqs. (8.15) and (8.16) imposes stability constraints on F, the free energy analog, and by inference, on the resource rate index Z driving it.

8.6 An Example

We again approximate the sum in Eq. (8.11) by an integral—so that $h(g(Z)) = g(Z)$—and make a simple assumption on the form of $dZ/dt = f(Z)$, say $f(Z) = \beta - \alpha Z(t)$. Then

$$g(Z) = -\frac{2\ln(Z)\,Z\beta - Z^2\alpha + 2C_1 Z - 2Z\beta + 2C_2}{2W_L\left(n, -\ln(Z)\,Z\beta + \frac{Z^2\alpha}{2} - C_1 Z + Z\beta - C_2\right)} \tag{8.17}$$

with, again, $L = \exp[-H_0/g(Z)]$, depending on the rate parameters α and β, the boundary conditions C_i, and the degree of the Lambert W-function. Proper choice of boundary conditions generates a classic 'inverted-U' signal transduction Yerkes-Dodson Law analog (e.g., Wallace 2020a, 2021c; Diamond et al. 2007). That is, since $Z \to \beta/\alpha$, we look at the cognition rate for fixed α and boundary conditions C_j as β increases. The result is shown in Fig. 8.5, for appropriate boundary conditions.

Similar results follow if

$$\exp[-F/g(Z)] = h(g(Z)) \propto A_m g(Z)^m$$

that is, if the function $h(g(Z))$ has a strongly dominant term of order $m > 0$.

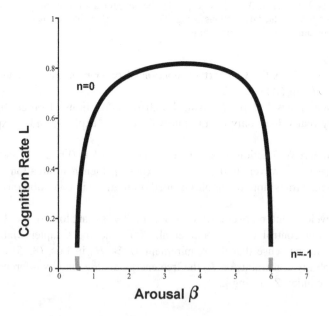

Fig. 8.5 Classic 'inverted-U' signal transduction for the cognition rate based on Eq. (8.17), setting $\alpha = 1$, $C_1 = -2$, $C_2 = -2$, $H_0 = 1$. Increase in β is taken as the 'arousal' measure

It is complicated—but not difficult—to incorporate stochastic effects into cognition rate dynamics based on Eq. (8.17), via standard methods from the theory of stochastic differential equations (e.g., Wallace 2021a,b,c). See the Chapter Mathematical Appendix for an example based on Fig. 8.5, fixing $\beta = 3$.

8.7 Cooperation: Multiple Workspaces

Individual brains—and indeed, even individual cells—are composed of interacting (often spatially distributed) cognitive submodules. Social groups are constituted by interacting individuals, separated by space, time, and/or social distance. Institutions are made up of dispersed but interacting 'workgroups', in a large sense. A critical phenomenon in all such examples is that the joint uncertainty of the dual information source associated with the particular level of cognition is less than or equal to the sum of the uncertainties of independent components—the information theory chain rule (Cover and Thomas 2006). To invert the argument, preventing crosstalk between cognitive submodules requires more investment of free energy or other resources than allowing interaction, as the electrical engineers often lament. From this lemon, evolution has made lemonade (e.g., Wallace 2022a).

More specifically for the work here, the emergence of a generalized Lambert W-function in Eq. (8.12), reducing to a 'simple' W-function if $h(g(Z)) = g(Z)$, is particularly suggestive. Recall that the fraction of network nodes in a giant component of a random network of N nodes with probability P of linkage between them can be given as (Newman 2010)

$$\frac{W_L(0, -NP \exp[-NP]) + NP}{NP} \tag{8.18}$$

where, again, the Lambert W-function emerges.

This expression has punctuated onset of a giant component of linked nodes only for $NP > 1$. See Fig. 8.6. In general, we might expect P to be a monotonic increasing function of $g(Z)$.

Within broadly 'social' groupings, interacting cognitive submodules—individuals—can become linked into shifting, tunable, temporary, workgroup equivalence classes to address similarly rapidly shifting patterns of threat and opportunity. These might range from complicated but relatively slow multiple global workspace processes of gene expression and immune function to the rapid—hence necessarily stripped-down—single-workspace neural phenomena of higher animal consciousness (Wallace 2012).

More complicated approaches to such a phase transition—involving Kadanoff renormalizations of the Morse Function free energy measure F and related measures—can be found in Wallace (2005, 2012, 2022a).

By contrast here, while multiple workspaces are most simply invoked in terms of a simultaneous set of the $g_j(Z_j)$, individual workspace tunability emerges from exploring equivalence classes of network topologies associated with a particular value of some designated $g_j(Z_j)$. Central matters then revolve around the equivalence class decompositions implied by the existence of the resulting workgroups, leading again to dynamic groupoid symmetry-breaking as they shift form and function in response to changing patterns of threat and opportunity. See Wallace (2021a, Sec. 12) for a parallel argument from an ergodic system perspective. Here, the $g(Z)$ substitute for the 'renormalization constant' ω in that development. For brevity, we omit a full discussion here.

8.8 Network Topology Is Important

We can, indeed, calculate the cognition rate of a single 'global workspace' as follows.

Suppose we have N linked cognitive submodules—individuals—in a social 'giant component', i.e., we operate in the upper portion of Fig. 8.6. The free energy Morse Function F can be expressed in terms of a full-bore partition function as

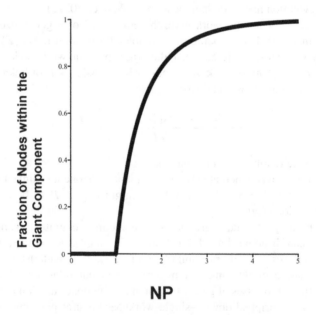

Fig. 8.6 Fraction of an N-node random network within the giant component as determined by the probability of contact between nodes P. The essential point is the punctuated accession to 'global broadcast' if and only if $NP > 1$ (e.g., Baars 1989; Dehaene and Changeux 2011). We might expect P to be a monotonic increasing function of $g(Z)$

$$\exp[-F/g(Z)] = \sum_{k=1}^{N} \sum_{j=1}^{M} \exp[-H_{k,j}/g(Z)]$$

$$\approx \sum_{k=1}^{N} \int_{0}^{\infty} \exp[-H_k/g(Z)]d H_k = \sum_{k=1}^{N} g(Z) = Ng(Z)$$

$$F = -\log[Ng(Z)]g(Z), \; g(Z) = \frac{-F}{W_L(n, -NF)} \qquad (8.19)$$

The sum over j represents available states within individual submodules, and the sum over k is across submodules, and W_L is again the Lambert W-function. Note, however, the appearance of the factor NF in the last expression.

The rate of cognition of the linked-node giant component can be expressed as

$$L = \frac{\sum_{k=1}^{N} \int_{H_0^k}^{\infty} \exp[-H_k/g(Z)]d H_k}{\sum_{k=1}^{N} \int_{0}^{\infty} \exp[-H_k/g(Z)]d H_k}$$

$$= \frac{g(Z) \sum_k \exp[-H_0^k/g(Z)]}{Ng(Z)}$$

$$= \left(\sum_k \exp[-H_0^k/g(Z)] \right) /N \equiv < L_k > \qquad (8.20)$$

where, not entirely unexpectedly, $< L_k >$ represents an averaging operation. More sophisticated averages might well be applied—at the expense of more formalism.

If we impose the approximation of Onsager nonequilibrium thermodynamics, i.e., defining $S(Z) = -F(Z) + ZdF/dZ$, and assume $dS/dZ = f(Z)$, we again obtain $f(Z) = Zd^2F/dZ^2$, and can calculate $g(Z)$ from the last expression in Eq. (8.19).

The appearance of N in expressions for $g(Z)$ and L is of some note. In particular, groupthink—high values of H_0^k—may result in failure to detect important signals.

Such considerations lead to a fundamentally different picture that is not just possible, but often observed, i.e., a transmission model. Consider something like a colony of prairie dogs under predation by hawks. A 'sentinel' pattern emerges, rather than the average system of Eq. (8.20), through the 'bottleneck' of a single, very highly-optimized, 'subcomponent' of the larger social structure. This involves a hypervigilant individual—or a small number of such appropriately dispersed— whose special danger signal or signals can be imposed rapidly across the entire population—postures, vocalizations, or both. Then

$$L = \max_k \{L_k = \exp[-H_0^k/g(Z)]\} \qquad (8.21)$$

where max is the maximization across the set $\{L_k\}$. However, Dore et al. (2019) show that, while brain activity in humans can track information sharing, there are likely to be important individual differences. That is, not everyone may respond the same to an 'alert' message.

Other network topologies are clearly possible. For example, in a rigidly hierarchical linear-chain business, military, or political setting, L will often be given by the *minimization* of cognition rates across the L_k, that is, by a bottleneck model.

The critical dependence of a system's cognitive dynamics on its underlying 'social topology' has profound implications for theories of the 'extended conscious mind' (Clark 2009; Lucia Valencia and Froese 2020). A particular expression of such matters involves failures of institutional cognition on wickedly hard problems (Wallace 2021b).

8.9 Time and Resource Constraints Are Important

Multiple workspaces, however, can also present a singular—and independent—problem of resource delivery, taking the scalar Z_j and *time itself* as essential resources. That is, not only are the Z_j both limited and delayed, in most cases, there will be a limit on possible response times, for both 'predator' and 'prey', so to speak.

If we assume an overall limit to available resources across a multiple workspace system $j = 1, 2, \ldots$ as $Z = \sum_j Z_j$, and available time as $T = \sum_j T_j$, then it is possible to carry out a simple Lagrangian optimization on the rate of system cognition $\sim \sum_j \exp[-H_j^0/g_j(Z_j)]$ as

$$\mathscr{L} \equiv \sum_j \exp[-H_j^0/g_j(Z_j)] +$$

$$\lambda\left(Z - \sum_j Z_j\right) + \mu\left(T - \sum_j T_j\right)$$

$$\partial\mathscr{L}/\partial Z_j = 0, \quad \partial\mathscr{L}/\partial T_j = 0 \tag{8.22}$$

where we assume $dZ_j/dt = f_j(Z_j(t)) = \beta_j - \alpha_j Z_j(t)$.

This leads, after some development, to a necessary expression for individual subsystem resource rates as

$$Z_j = f_j^{-1}(\mu/\lambda) = \frac{\beta_j - \mu/\lambda}{\alpha_j} > 0 \tag{8.23}$$

μ and λ are to be viewed in economic terms as *shadow prices* imposed by 'environmental constraints', in a large sense (e.g., Jin et al. 2007; Robinson 1993).

The essential point is that cognitive dynamics—the rates of cognition driven by the rate of available resources Z_j and time—in this model, are strongly determined by the shadow price ratio μ/λ. The shadow price ratio is to be interpreted as an environmental signal. A sufficiently large shadow price ratio, according to Eq. (8.23), can starve essential components, driving their cognition rates to failure.

8.10 Further Theoretical Development

Following the arguments of Wallace (2021a), the cognition models can be extended in a number of possible directions, much as is true for 'ordinary' regression theory.

Perhaps the simplest next step is to replace the relation $dZ/dt = f(Z(t))$ with a stochastic differential equation having the form $dZ_t = f(Z_t)dt + \sigma g(Z_t)dB_t$, where dB_t represents ordinary white noise.

A next 'simple' generalization might be replacing the scalar index Z with a multidimensional vector quantity, $\mathbf{Z}$, leading to an intricate set of simultaneous partial differential equations requiring, at best, Lie symmetry address. See the Mathematical Appendix for an outline.

Further development could involve expanding the 'Onsager approximation' in terms of a 'generalized entropy' $S = \sum_k \epsilon_k Z^{k-1} F^{k-1}$, where $F^j \equiv d^j F/dZ^j$. Then the dynamics might also be generalized, at least for a scalar Z, as $\partial Z/\partial t \approx \sum_j \mu_j d^j S/dZ^j = f(Z)$, and so on toward multidimensional models.

As with regression equations much beyond $Y = mX + b$, matters can rapidly become complicated indeed.

8.11 Discussion

Hasson et al. (2012), in a widely-cited foundational study, call for a reorientation of neuroscience from a single-brain to a multi-brain frame of reference:

> Cognition materializes in an interpersonal space. The emergence of complex behaviors requires the coordination of actions among individuals according to a shared set of rules. Despite the central role of other individuals in shaping our minds, most cognitive studies focus on processes that occur within a single individual. We call for a shift from a single-brain to a multi-brain frame of reference. We argue that in many cases the neural processes in one brain are coupled to the neural processes in another brain via the transmission of a signal through the environment. Brain-to-brain coupling constrains and simplifies the actions of each individual in a social network, leading to complex joint behaviors that could not have emerged in isolation.

Kingsbury et al. (2019), Rose et al. (2021) and Baez-Mendoza et al. (2021) extend this perspective to interacting non-human populations, while Abraham et al. (2020), for humans, extend the time scale across generations.

Here, we outline something of the formal developments needed to implement such reorientation, based on groupoid symmetry-breaking within longstanding paradigms of information and control theories, as affected and afflicted by network topologies and their dynamics. This, it can be argued, is very much a 'rocket science' problem, since the difficulty lies not in the individual components of a possible comprehensive approach, which are all well-studied, but in using the building blocks to construct a theoretical enterprise that accounts well for observational and experimental data. Very similar conundrums confront contemporary theories of consciousness (e.g., Wallace 2022a), albeit, in this case, without the hindrance of quite so many longstanding philosophical and de-facto theological presuppositions.

Application of the approach developed here to stochastic systems is straightforward if somewhat complicated, as is extension to higher approximation 'Onsager-type' entropy gradient models, the analog of moving from $y = mx + b$ to $y = mx^2 + b$, and so on (Wallace 2021a,b).

In a deep-time sense, the underlying mechanisms have long been with us, i.e., the evolutionary exaptation of the inevitable second-law 'leakage' of crosstalk between co-resident cognitive processes (e.g., Wallace 2012). Crosstalk characterizes the immune system, wound-healing, tumor control, gene expression, and so on, up through and including far more rapid neural processes. It is not a great leap-of-faith to infer that similar dynamics instantiate social interactions between individuals within and across populations.

8.12   Mathematical Appendix

Groupoids

Following Brown (1992), consider a directed line segment in one component, written as the source on the left and the target on the right.

$$\bullet \longrightarrow \bullet$$

Two such arrows can be composed to give a product **ab** if and only if the target of **a** is the same as the source of **b**

$$\bullet \xrightarrow{\ a\ } \bullet \xrightarrow{\ b\ } \bullet$$

Brown puts it this way,

One imposes the geometrically obvious notions of associativity, left and right identities, and inverses. Thus a groupoid is often thought of as a group with many identities, and the reason why this is possible is that the product **ab** is not always defined.

We now know that this apparently anodyne relaxation of the rules has profound consequences... [since] the algebraic structure of product is here linked to a geometric structure, namely that of arrows with source and target, which mathematicians call a directed graph.

Cayron (2006) elaborates:

A group defines a structure of actions without explicitly presenting the objects on which these actions are applied. Indeed, the actions of the group G applied to the identity element e implicitly define the objects of the set G by ge = g; in other terms, in a group, actions and objects are two isomorphic entities. A groupoid enlarges the notion of group by explicitly introducing, in addition to the actions, the objects on which the actions are applied. By this approach, many identities may exist (they correspond to the actions that leave an object invariant).

It is of particular importance that equivalence class decompositions permit construction of groupoids in a highly natural manner.

Weinstein (1996) and Golubitsky and Stewart (2006) provide more details on groupoids and on the relation between groupoids and bifurcations.

An essential point is that, since there are no necessary products between groupoid elements, 'orbits', in the usual sense, disjointly partition groupoids into 'transitive' subcomponents.

The Data Rate Theorem

Real-world environments are inherently unstable. Organisms, to survive, must exert a considerable measure of control over them. These control efforts range from immediate responses to changing patterns of threat and affordance, through niche construction, and, in higher animals, elaborate, highly persistent, social and sociocultural structures. Such necessity of control can, in some measure, be represented by a powerful asymptotic limit theorem of probability theory different from, but as fundamental as, the Central Limit Theorem: the Data Rate Theorem, first derived as an extension of the Bode Integral Theorem of signal theory.

Consider a reduced model of a control system as follows:

For the inherently unstable system of Fig. 8.7, assume an initial n-dimensional vector of system parameters at time t, as x_t. The system state at time $t + 1$ is then—near a presumed nonequilibrium steady state—determined by the first-order relation

$$x_{t+1} = \mathbf{A}x_t + \mathbf{B}u_t + W_t \tag{8.24}$$

In this approximation, $\mathbf{A}$ and $\mathbf{B}$ are taken as fixed n-dimensional square matrices. u_t is a vector of control information, and W_t is an n-dimensional vector of Brownian white noise.

According to the DRT, if H is a rate of control information sufficient to stabilize an inherently unstable control system, then it must be greater than a minimum, H_0,

Fig. 8.7 The reduced model
of an inherently unstable
system stabilized by a control
signal U_t

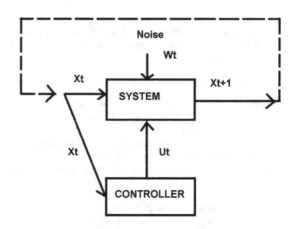

$$H > H_0 \equiv \log[\| \det[\mathbf{A}^m] \|] \tag{8.25}$$

where det is the determinant of the subcomponent $\mathbf{A}^m$—with $m \leq n$—of the matrix $\mathbf{A}$ having eigenvalues ≥ 1. H_0 is defined as the rate at which the unstable system generates 'topological information' on its own.

If this inequality is violated, stability fails.

Stochastic Analysis for Fig. 8.5

Here, we apply the Ito Chain Rule (Protter 2005) to the expression $L(Z) = \exp[-1/g(Z)]$, as based on Eq. (8.17) for $g(Z)$. We set $\alpha = 1$, $\beta = 3$, $C_1 = C_2 = -2$ and numerically calculate the solution set for the relation $< dL_t >= 0$ based on the underlying stochastic differential equation

$$dZ_t = (\beta - \alpha Z_t)dt + \sigma Z_t dB_t \tag{8.26}$$

where dB_t is assumed to be ordinary white noise, as associated with Brownian motion.

Figure 8.8 shows the result, the solution equivalence class $\{\sigma, Z\}$ for $\beta = 3$, just to the left of the peak in Fig. 8.5. Note that, at $\sigma \approx 0.278$, the system becomes susceptible to a bifurcation instability, well before the 'standard' instability expected from a second-order Ito Chain Rule analysis based on Eq. (8.26). More specifically, the nonequilibrium steady state (nss) conditions associated with Eq. (8.26) are the relations $< Z_t >= \beta/\alpha$ and $\mathrm{Var}[Z_t] = \left(\beta/(\alpha - \sigma^2/2)\right)^2 - (\beta/\alpha)^2$, so that variance in Z explodes as $\sigma \to \sqrt{2\alpha}$, here, $\sqrt{2}$.

Similar analyses to Fig. 8.8 across increasing values of β produce increasingly complicated equivalence classes $\{\sigma, Z\}$, as constrained by the nss conditions on Z_t.

Fig. 8.8 Numerical solution equivalence class {σ, Z} for the relation $< dL_t > = 0$ from figure 8.5, taking $\beta = 3$, just to the left of the peak. Again, $\alpha = 1$, $C_1 = C_2 = -2$. While, in this model, Z_t becomes unstable in variance at $\sigma > \sqrt{2}$, the cognition rate can suffer a bifurcation instability for $\sigma > \approx 0.278$

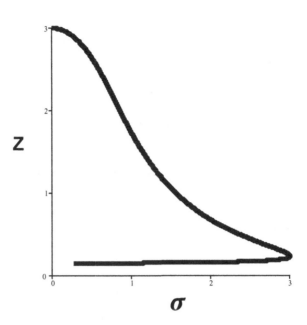

Higher Dimensional Systems

Above, we have viewed systems as sufficiently well characterized by the single scalar parameter Z, mixing material resource/energy supply with internal and external flows of information. The real world, however, may often be far more complicated. That is, invoking techniques akin to Principal Component Analysis, there may be more than one independent composite entity irreducibly driving system dynamics. It may then be necessary to replace the scalar Z with an n-dimensional vector $\mathbf{Z}$ having orthogonal components that, together, account for a good portion of the total variance in the rate of supply of essential resources. The dynamic equations are then in vector form:

$$F(\mathbf{Z}) = -\log(h(g(\mathbf{Z}))) g(\mathbf{Z})$$

$$S = -F + \mathbf{Z} \cdot \nabla_{\mathbf{Z}} F$$

$$\partial \mathbf{Z}/\partial t \approx \hat{\mu} \cdot \nabla_{\mathbf{Z}} S = f(\mathbf{Z})$$

$$-\nabla_{\mathbf{Z}} F + \nabla_{\mathbf{Z}}(\mathbf{Z} \cdot \nabla_{\mathbf{Z}} F) =$$

$$\hat{\mu}^{-1} \cdot f(\mathbf{Z}) \equiv f^*(\mathbf{Z})$$

$$\left(\left(\partial^2 F / \partial z_i \partial z_j \right) \right) \cdot \mathbf{Z} = f^*(\mathbf{Z})$$

$$\left(\left(\partial^2 F / \partial z_i \partial z_j \right) \right) |_{\mathbf{Z}_{nss}} \cdot \mathbf{Z}_{\mathbf{nss}} = \mathbf{0} \qquad (8.27)$$

Here, F, g, h, and S are scalar functions, and $\hat{\mu}$ is an n-dimensional square matrix of diffusion coefficients. The matrix $((\partial F/\partial z_i \partial z_j))$ is the obvious n-dimensional square matrix of second partial derivatives, and $f(\mathbf{Z})$ is a vector function. The last relation imposes a nonequilibrium steady state condition, i.e. $f^*(\mathbf{Z}_{nss}) = \mathbf{0}$.

For the 'simple' Rate Distortion approach, $h(g(\mathbf{Z})) \rightarrow g(\mathbf{Z})$, while, again, we assume $\mathbf{Z}(t) \rightarrow \mathbf{Z}_{nss}$.

For $n \geq 2$, this is an overdetermined system of partial differential equations (Spencer 1969). Indeed, for a general $f^*(\mathbf{Z})$ the system is inconsistent, resulting in as many as n different expressions for $F(\mathbf{Z})$, and hence the same number of 'temperature' measures as determined by the relation $F = -\log(g)g$.

References

Abraham, E., J. Posner, P. Wickramaratne, N. Aw, M. van Dijk, J. Cha, M. Weissman, and A. Talati. 2020. Concordance in parent and offspring cortico-basal ganglia white matter connectivity varies by parental history of major depressive disorder and early parental care. Social Cognitive and Affective Neuroscience 15:889–903. https://doi.org/10.1093/scan/nsaa118

Atlan H., and I. Cohen. 1998. Immune information, self-organization, and meaning. International Immunology 10:711–717.

Baars B. 1989. *A Cognitive Theory of Consciousness*. New York: Cambridge University Press.

Baez-Mendoza, R., E. Mastrobattista, A. Wang, and Z. Williams. 2021. Social agent identity cells in the prefrontal cortex of interacting groups of primates. Science 374:421. https://doi.org/10.1126/science.abb4149

Barraza, P., A. Perez, and E. Rodriguez. 2020. Brain-to-brain coupling in the gamma-band as a marker of shared intentionality. Frontiers in Human Neuroscience 14:295. https://doi.org/10.3389/fnhum.2020.00295

Brown, R. 1992. Out of line. Royal Institute Proceedings 64:207–243.

Cayron, C. 2006. Groupoid of orientational variants. Acta Crystalographica Section A A62:21040.

Champagnat, N., R. Ferriere, and S. Meleard. 2006. Unifying evolutionary dynamics: from individual stochastic process to macroscopic models. Theoretical Population Biology 69:297–321.

Clark A. 2009. Spreading the joy? Why the machinery of consciousness is (probably) still in the head. Mind 118:963–993.

Cover, T., and J. Thomas. 2006. *Elements of Information Theory*, 2nd ed. New York: Wiley.

Dashorst, P., T. Mooren, R. Kleber, P. de Jong, and R. Huntjens. 2019. Intergenerational consequences of the Holocaust on offspring mental health: a systematic review of associated factors and mechanisms. European Journal of Psychotraumatology 10:1654065. https://doi.org/10.1080/20008198.2019.1654065

de Groot, S., and P. Mazur. 1984. *Nonequilibrium Thermodynamics*. New York: Dover.

Dehaene, S., and J. Changeux. 2011. Experimental and theoretical approaches to conscious processing. Neuron 70:200–227.

Dembo, A., and O. Zeitouni. 1998. *Large Deviations and Applications*, 2nd ed. New York: Springer.

Diamond, D., Campbell A., Park C., Halonen J., and Zoladz P. 2007. The temporal dynamics model of emotional memory processing. Neural Plasticity 2007:60803. https://doi.org/10.1155/2007/60803

Dikker, S., et al. 2017. Brain-to-brain synchrony tracks real-world dynamic group interactions in the classroom. Current Biology 27:1375–1380. https://doi.org/10.1016/j.cub.2017.04.002

Dolan, B., W. Janke, D. Johnston, and M. Stathakopoulos. 2001. Thin Fisher zeros. Journal of Physics A 34:6211–6223.

Dore, B., C. Scholz, E. Baek, J. Garcia, M. O'Donnell, D. Bassett, J. Vettel, and E. Falk. 2019. Brain activity tracks population information sharing by capturing consensus judgements of value. Cerebral Cortex 29:3102–3110. https://doi.org/10.1093/cercor/bhy176

Dretske, F. 1994. The explanatory role of information. Philosophical Transactions of the Royal Society A 349:59–70.

Feynman, R. 2000. *Lectures in Computation*. Boulder: Westview Press.

Fisher, M. 1965. *Lectures in Theoretical Physics*, vol. 7. Boulder: University of Colorado Press.

Golubitsky, M., and I. Stewart. 2006. Nonlinear dynamics and networks: the groupoid formalism. Bulletin of the American Mathematical Society 43:305–364.

Hari, R., L. Henriksson, S. Malinen, and L. Parkkonen. 2015. Centrality of social interaction in human brain function. Neuron 88:181–193.

Hasson, U., and C. Frith. 2016. Mirroring and beyond: coupled dynamics as a generalized framework for modelling social interactions. Philosophical Transactions B 371:20150366. http://dx.doi.org/10.1098/rstb.2015.0366

Hasson, U., A.A. Ghazanfar, B. Galantucci, S. Garrod, and C. Keysers. 2012. Brain-to-brain coupling: a mechanism for creating and sharing a social world. Trends in Cognitive Science 16:114–121. https://doi.org/10.1016/j.tics.2011.12.007

Henry, T., D. Robinaugh, and E. Fried. 2022. On the control of psychological networks. Psychometrika 87:188–213. https://doi.org/10.1007/s11336-021-09796-9

Hoyrup, M. 2013. Computability of the ergodic decomposition. Annals of Pure and Applied Logic 164:542–549.

Jin, H., Z. Hu, and X. Zhou. 2008. A convex stochastic optimization problem arising from portfolio selection. Mathematical Finance 18:171–183.

Khinchin, A. 1957. *Mathematical Foundations of Information Theory*. New York: Dover.

Kingsbury, L., and W. Hong. 2020. A multi-brain framework for social interaction. Trends in Neurosciences 43:651–665. https://doi.org/10.1016/j.tins.2020.06.008

Kingsbury, L., S. Huang, J. Wang, K. Gu, P. Golshani, Y. Wu, and W. Hong. 2019. Correlated neural activity and encoding of behavior across brains of socially interacting animals. Cell 178:429–446. https://doi.org/10.1016/j.cell.2019.05.022

Kuhlen, A., C. Bogler, S.E. Brennan, and J.D. Haynes. 2017. Brains in dialogue: decoding neural preparation of speaking to a conversational partner. Social Cognitive and Affective Neuroscience 12: 871–880. https://doi.org/10.1093/scan/nsx018

Laidler, K. 1987. *Chemical Kinetics*, 3rd ed. New York: Harper and Row.

Lucia Valencia, A., and T. Froese. 2020. What binds us? Inter-brain neural synchronization and its implications for theories of human consciousness. Neuroscience of Consciousness 6:niaa010. https://doi.org/10.1093/nc/niaa010

Maturana, H., and F. Varela. 1980. *Autopoiesis and Cognition: The Realization of the Living*. Boston: Reidel.

Mezo I., and G. Keady. 2015. Some physical applications of generalized Lambert functions. arXiv:1505.01555v2 [math.CA] 22 Jun 2015.

Nair, G., F. Fagnani, S. Zampieri, and R. Evans. 2007. Feedback control under data rate constraints: an overview. Proceedings of the IEEE 95:108137.

Newman, M. 2010. *Networks: An Introduction*. New York: Oxford University Press.

Perez, A., M. Carreiras, and J.A. Duñabeitia. 2017. Brain-to-brain entrainment: EEG interbrain synchronization while speaking and listening. Scientific Reports 7:4190. https://doi.org/10.1038/s41598-017-04464-4

Pettini, M. 2007. *Geometry and Topology in Hamiltonian Dynamics and Statistical Mechanics*. New York: Springer.

Protter, P. 2005. *Stochastic Integration and Differential Equations*, 2nd ed. New York: Springer.

Robinson, S. 1993. Shadow prices for measures of effectiveness II: general model. Operations Research 41:536–548.

Rose, M., B. Styr, T. Schmid, J. Elie, and M. Yartsev. 2021. Cortical representation of group social communication in bats. Science 374:422. https://doi.org/10.1126/science.aba9584

Ruelle, D. 1964. Cluster property of the correlation functions of classical gases. Reviews of Modern Physics 36:580–584.

Shields, P., D. Neuhoff, L. Davisson, and F. Ledrappier. 1978. The distortion-rate function for nonergodic sources. Annals of Probability 6:138–143.

Sliwa, J. 2021. Toward collective animal neuroscience. Science 374:397–398.

Spencer, D. 1969. Overdetermined systems of linear partial differential equations. Bulletin of the American Maththematical Society 75:179–239.

Stolk, A., et al., 2014. Cerebral coherence between communicators marks the emergence of meaning. Proceedings of the National Academy of Sciences USA 111:18183–18188.

Wallace, R. 2005. *Consciousness: A Mathematical Treatment of the Global Neuronal Workspace Model*. New York: Springer.

Wallace, R. 2012. Consciousness, crosstalk, and the mereological fallacy: an evolutionary perspective. Physics of Life Reviews 9:426–453.

Wallace, R. 2018. New statistical models of nonergodic cognitive systems and their pathologies. Journal of Theoretical Biology 436:72–78.

Wallace, R. 2020a. On the variety of cognitive temperatures and their symmetry-breaking dynamics. Acta Biotheoretica 68:421–439. https://doi.org/10.1007/s10441-019-09375-7

Wallace, R. 2020b. *Cognitive Dynamics on Clausewitz Landscapes: The Control and Directed Evolution of Organized Conflict*. New York: Springer.

Wallace, R. 2021a. Toward a formal theory of embodied cognition. BioSystems 202:104356.

Wallace, R. 2021b. Enbidued cognition and its pathologies: the dynamics of institutional failure on wickedly hard problems. Communications in Nonlinear Science and Numerical Simulation 95:105616.

Wallace, R. 2021c. How AI founders on adversarial landscapes of fog and friction. Journal of Defense Modeling and Simulation https://doi.org/10.1177/1548512920962227

Wallace, R. 2022a. Consciousness, *Cognition and Crosstalk: The Evolutionary Exaptation of Nonergodic Groupoid Symmetry-Breaking*. New York: Springer.

Wallace, R. 2022b. Major transitions as groupoid symmetry-breaking in nonergodic prebiotic, biological and social information systems. Acta Biotheoretica 20:27. https://doi.org/10.1007/s10441-022-09451-5

Weinstein, A. 1996. Groupoids: unifying internal and external symmetry. Notices of the American Mathematical Association 43:744–752.

Yeshurun, Y., M. Nguyen, and U. Hasson. 2021. The default mode network: where the idiosyncratic self meets the shared social world. Nature Reviews Neuroscience 22:181–192. https://www.nature.com/articles/s41583-020-00420-w

Yi, S., P.W. Nelson, and A.G. Ulsoy. 2010. *Time-Delay Systems: Analysis and Control Using the Lambert W Function*. New Jersey: World Scientific.

Chapter 9
Afterward

Perhaps the most defining characteristic of the living state is the centrality of cognition at every scale and level of organization, in the embedding context of interaction, contention, and selection—ubiquitous evolutionary processes at and across those same scales and levels of organization (e.g., Maturana and Varela 1980; Atlan and Cohen 1998; Pigliucce and Muller 2010). Since cognitive phenomena take place on highly dynamic 'roadways' of imprecision and uncertainty in both threat and affordance, they are inherently unstable. Since cognition implies choice of action, and choice implies reduction of uncertainty, the asymptotic limit theorems of control and information theories serve as primary tools for the address of cognition on the landscapes of instability and selection that are the playing fields of evolutionary dynamics. Hence the formal developments in this collection.

By virtue of the challenges of inherent instability, cognitive phenomena are most often closely paired with regulatory 'riverbanks', and the dysfunctions of regulation underlie many of the pathologies associated with developmental disorders, chronic disease, mental illness, and aging. Widespread outbreaks of pandemic infection in modern states can be seen as a failure of institutional cognition (e.g., Wallace 2022). Indeed, such failure also drives large-scale loss of life consequent on strategic incompetence in armed conflict (Krepinevich and Watts 2009a,b; Wallace 2020). That is, 'cognitive failure', in a large sense, underlies much of the impact of embedding selection pressures driving evolution at and across the various modes of biological, social, and institutional enterprise.

Although we do not emphasize the matter here, automata, of varying degrees of sophistication, while not cognitive at every scale and level of organization, nonetheless share some—increasingly many—of the information and control theory constraints limiting all processes of cognition and deliberative action under uncertainty, adversarial intent, and other selection pressures (e.g., Wallace 2019, 2021).

This volume brings cognitive, regulatory, and evolutionary phenomena under a broad theoretical framework of probability models based on, but not limited to, understanding gained from information theory, control theory, statistical physics,

R. Wallace, *Essays on the Extended Evolutionary Synthesis*, SpringerBriefs in Evolutionary Biology, https://doi.org/10.1007/978-3-031-29879-0_9

and nonequilibrium thermodynamics. In addition to providing significant scientific insight on the nature of evolutionary process and its manifold expressions, the construction of such probability models is a first step toward building robust statistical tools to aid in the analysis of observational and experimental data, the only sure basis of new knowledge as opposed to new theoretical speculation. Turning probability models into useful statistical tools is no easy thing, and largely remains to be done for the work presented here. The scientific, technological, and social benefits of such an effort, however, are likely to be considerable.

References

Atlan H., and I. Cohen. 1998. Immune information, self-organization, and meaning. International Immunology 10:711–717.

Krepinevich, A., and B. Watts. 2009a. *Retaining Strategic Competence*. Washington D.C.: Center for Strategic and Budgetary Assessment.

Krepinevich, A., and B. Watts. 2009b. Lost at the NSC. The National Interest 99:63–72.

Maturana, H., and F. Varela. 1980. *Autopoiesis and Cognition: The Realization of the Living*. Boston: Reidel.

Pigliucce, M., and G. Muller. 2010. *Evolution: The Extended Synthesis*. Cambridge: MIT Press.

Wallace, R. 2019. Cognitive instabilities under contention, friction, and the fog-of-war challenge the AI revolution. Connection Science 32:264–279. https://doi.org/10.1080/09540091.2019.1684441

Wallace, R. 2020. *Cognitive Dynamics on Clausewitz Landscapes: The Control and Directed Evolution of Organized Conflict*. New York: Springer.

Wallace, R. 2021. How AI founders on adversarial landscapes of fog and friction. Journal of Defense Modeling and Simulation 19. https://doi.org/10.1177/1548512920962227

Wallace, R. 2022. Essays on Strategy and Public Health: The Systematic Reconfiguration of Power Relations. New York: Springer.

Printed in the United States
by Baker & Taylor Publisher Services